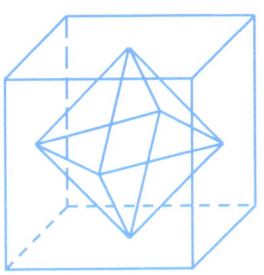  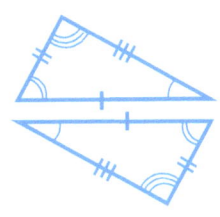 

A Basic Course in Geometry

– Part 2 of 5

Bill Lembke

Citrus Software Publishing

Citrus Ridge, Florida

Published by Citrus Software Publishing, a Division of Citrus Software Corporation

Copyright © 2013 by Citrus Software Corporation

All rights reserved. No part of this publication may be reproduced or distributed in any form or by any means, or stored in a database or retrieval system, without the prior written consent of the publisher, including, but not limited to, network storage or transmission, or broadcast for distance learning.

Printed and bound in the United States of America.

Citrus Software, Citrus Software Publishing, and ABC Method of Instruction are either registered trademarks or trademarks of Citrus Software Corporation in the United Stated and/or other countries. Other products and company names mentioned herein may be the trademarks of their respective owners.

This book expresses the author's views and opinions. The information contained in this book is provided without any express, statutory, or implied warranties. Neither the author, Citrus Software Corporation, nor its resellers, or distributors will be held liable for any damages caused or alleged to be caused either directly or indirectly by this book.

A Basic Course in Geometry – Part 2 of 5

ISBN-13: 978-1477557297

ISBN-10: 1477557296

Table of Contents – Part 2 of 5

Chapter 3 – Polytopes   1

- 3-1 Introduction   1
- 3-2 Types of Polytopes   2
- 3-3 Regular Convex Polytopes   2
- 3-4 Polytopes – Simplex, Orthoplex, and Hypercube   4
- 3-5 Polytope Decomposition   7
- 3-6 Summary   8
- Chapter Test   10

Chapter 4 – Polygons   13

- 4-1 Introduction   13
- 4-2 Polygon Classification   13
- 4-3 Polygon Parts   16
- 4-4 Polygon Formulas   17
- 4-5 Polygon Names   17
- 4-6 Regular Polygon Properties   19
- 4-7 Triangle   20
- 4-8 Triangle Formulas   22
- 4-9 Trigonometry   24
- 4-10 Regular Polygon Areas   25
- 4-11 Drawing Regular Polygons   26
- 4-12 Summary   27
- Chapter Test   30

Chapter 5 – Triangles and Quadrilaterals   34

- Section 1 – Triangles   34
- 5-1 Introduction   34
- 5-2 Triangle Types   34
- 5-3 Similar and Congruent Triangles   38
- 5-4 Points, Lines, and Circles of Triangles   41
- 5-5 Triangle Rigidity   41

- 5-6 Summary   47
- Section 2 – Quadrilaterals   49
- 5-7 Introduction   49
- 5-8 Convex Quadrilaterals   49
- 5-9 Quadrilateral Formulas   51
- 5-10 Summary   54
- Chapter Test   55

Chapter 6 – Polyhedron   59

- 6-1 Introduction   59
- 6-2 Polyhedron Name Conventions   60
- 6-3 Polyhedron Name – Number of Faces   61
- 6-4 Polyhedron Name – Shape of Faces   62
- 6-5 Polyhedron Symmetry   63
- 6-6 Polyhedron Characteristics   65
- 6-7 Euler's Formula   67
- 6-8 Polyhedron Nets   68
- 6-9 Summary   69
- Chapter Test   71

CHAPTER 3: Polytopes

3-1 Introduction

A polytope is a geometric object defined as a finite region of n-dimensional space, where *n* is an arbitrary number, enclosed by a finite number of hyperplanes. It is a multidimensional solid with flat sides. A hyperplane is a generalization of the concept of a plane. In a one-dimensional space, such as a line, a hyperplane is a point; it divides a line into two rays. In two-dimensional space, such as the plane, a hyperplane is a line; it divides the plane into two half-planes. In three-dimensional space, a hyperplane is an ordinary plane; it divides the space into two half-spaces. This concept can also be applied to four-dimensional space and beyond, where the dividing object is simply referred to as a hyperplane. In the figures below, hyperplanes are shown in one-dimensional, two-dimensional, and three-dimensional space.

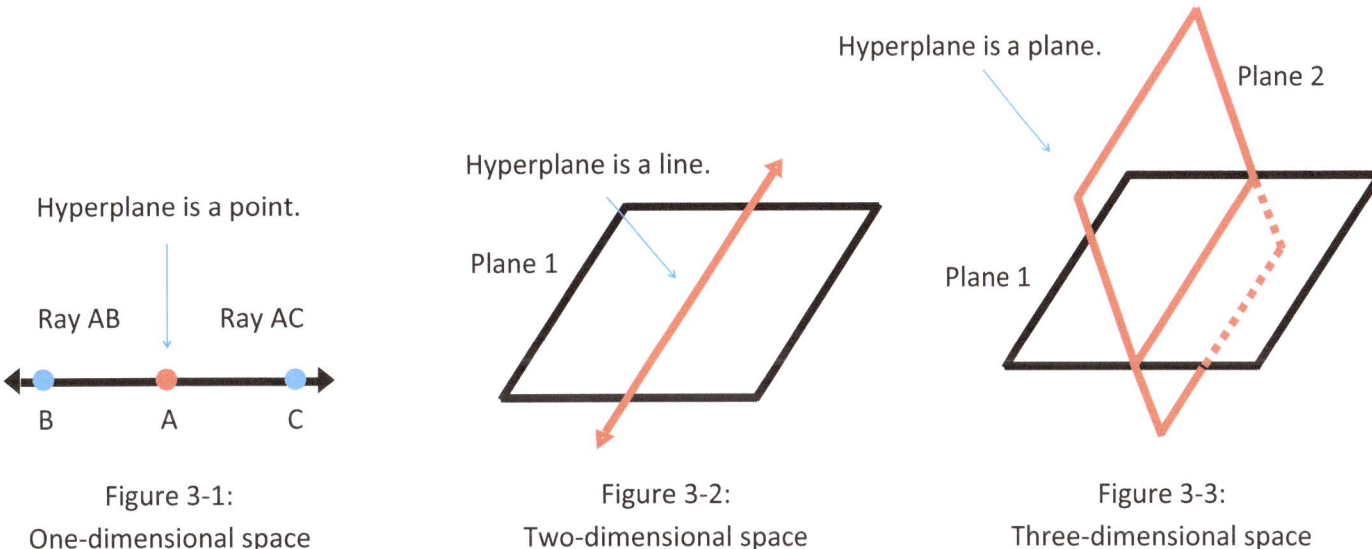

Figure 3-1: One-dimensional space

Figure 3-2: Two-dimensional space

Figure 3-3: Three-dimensional space

In zero-dimensional space, the only possible object, or polytope, is a point. It does not have a size, width, height, depth, area, or volume. In one-dimensional space, the only possible polytopes are points and line segments. A line segment is bounded by two end points. In two-dimensional space, there are points, line segments, and polygons. The simplest possible polygon is the triangle. A triangle is bounded by three line segments (the sides) and three vertices (the end points of the three sides). In three-dimensional space, a polyhedron is bounded by polygons, which in turn are bounded by line segments, which in turn are bounded by points. A polyhedron is bounded by faces (polygons), edges (line segments), and vertices (points). Every edge of a polyhedron is created by the joining of two faces, and every vertex of a polygon is created by the joining of two sides. In four-dimensional space, a polychoron is bound by cells (polyhedra).

3-2 Types of Polytopes

When referring to an *n*-dimensional generalization, the term *n*-polytope is used. For example, a polygon is a 2-polytope, a polyhedron is a 3-polytope, and a polychoron is a 4-polytope. The table below shows polytopes dimensions, names, and examples.

Dimension	Polytope Name	n-Dimension	Examples
0	Monad	0-tope	Point
1	Polytelon	1-tope	Line Segment
2	Polygon	2-tope	Triangle, Square, Pentagon
3	Polyhedron	3-tope	Tetrahedron, Cube, Octahedron
4	Polychoron	4-tope	Pentachoron, Hexadecachoron, Tesseract
5	Polyteron	5-tope	Hexateron, Pentacross, Penteract
6	Polypeton	6-tope	Heptapeton, Hexacross, Hexeract
7	Polyexon	7-tope	Octaexon, Heptacross, Hepteract
8	Polyzetton	8-tope	Enneazetton, Octacross, Octeract
9	Polyyotton	8-tope	Decayotton, Enneacross, Enneract
10	Polyxennon	10-tope	Hendecaxennon, Decacross, Dekeract

Table 3-1: Polytope dimensions

3-3 Regular Convex Polytopes

Regular convex polytopes are classified primarily according to their dimensionality. Three special classes of regular polytope exist in every dimensionality: (1) simplex, (2) orthoplex, and (3) hypercube.

A simplex is a generalization of a triangle to an arbitrary dimension. An *n*-simplex is an *n*-dimensional polytope with n + 1 vertices. It has the smallest number of sides needed to create a convex shape containing the given vertices, or the simplest possible shape for a dimension. The following table shows simplexes in order of dimensionality.

Dimension	Simplex
0	Point
1	Line Segment
2	Trigon (Triangle)
3	Tetrahedron
4	Pentachoron
5	Hexateron

Table 3-2: Polytope Simplexes

An orthoplex is a generalization of square to an arbitrary dimension. An *n*-orthoplex is an *n*-dimensional polytope where the vertices are in pairs orthogonal or perpendicular to each other. The following table shows orthoplexes in order of dimensionality.

Dimension	Orthoplex
0	Point
1	Line Segment
2	Tetragon (Square)
3	Octahedron
4	Hexadecachoron (Tessacross)
5	Triacontakaiditeron (Pentacross)

Table 3-3: Polytope Orthoplexes

A hypercube is a generalization of square to an arbitrary dimension. An *n*-cube is an *n*-dimensional polytope where the sides are in parallel pairs orthogonal or perpendicular to each other. The following table shows hypercubes in order of dimensionality.

Dimension	Hypercube
0	Point
1	Line Segment
2	Tetragon (Square)
3	Hexahedron (Cube)
4	Octachoron (Tesseract)
5	Decateron (Penteract)

Table 3-4: Polytope Hypercubes

3-4 Polytopes – Simplex, Orthoplex, and Hypercube

The three cases of polytopes are simplex, orthoplex, and hypercube. The figures below show examples of polytopes for dimensions 2 through 10 for each of these cases.

Convex regular 2-polytopes are below.

      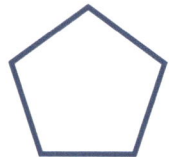

Figure 3-4: Triangle   Figure 3-5: Square   Figure 3-6: Pentagon
(3 sides)   (4 sides)   (5 sides)

Convex regular 3-polytopes are below.

Figure 3-7: Tetrahedron   Figure 3-8: Cube   Figure 3-9: Octahedron
(4 faces)   (6 faces)   (8 faces)

Convex regular 4-polytopes are below.

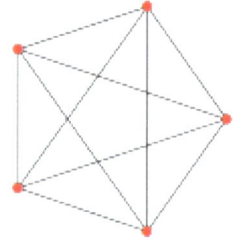   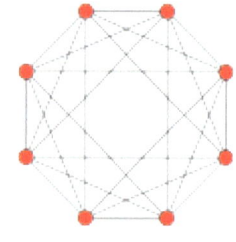   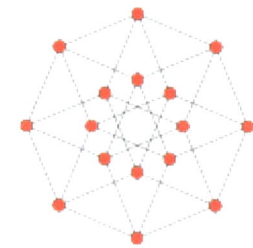

Figure 3-10: 4-simplex   Figure 3-11: 4-orthoplex   Figure 3-12: 4-cube
(Pentachoron)   (Hexadecachoron)   (Tesseract)

Convex regular 5-polytopes are below.

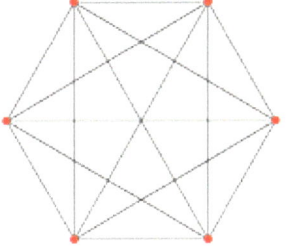
Figure 3-13: 5-simplex (Hexateron)

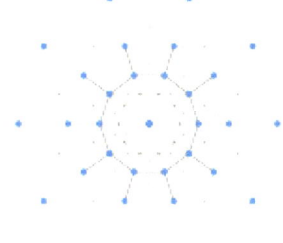
Figure 3-14: 5-orthoplex (Pentacross)

Figure 3-15: 5-cube (Penteract)

Convex regular 6-polytopes are below.

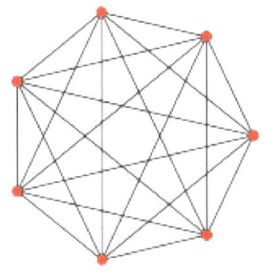
Figure 3-16: 6-simplex (Heptapeton)

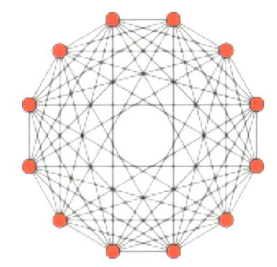
Figure 3-17: 6-orthoplex (Hexacross)

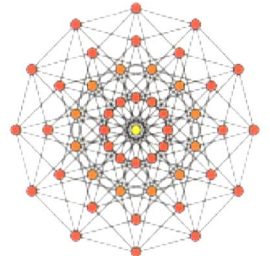
Figure 3-18: 6-cube (Hexeract)

Convex regular 7-polytopes are below.

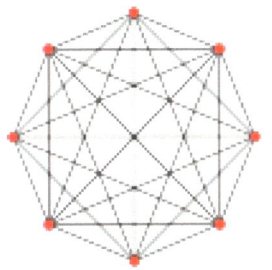
Figure 3-19: 7-simplex (Octaexon)

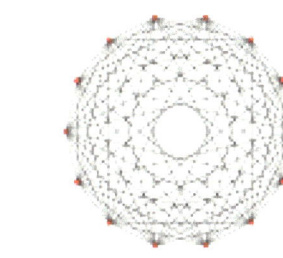
Figure 3-20: 7-orthoplex (Heptacross)

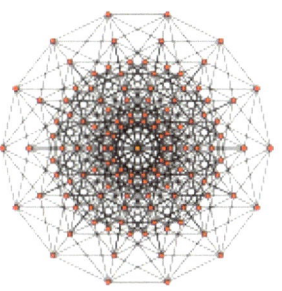
Figure 3-21: 7-cube (Hepteract)

Convex regular 8-polytopes are below.

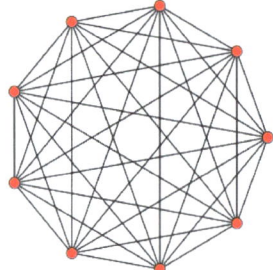

Figure 3-22: 8-simplex (Enneazetton)

Figure 3-23: 8-orthoplex (Octacross)

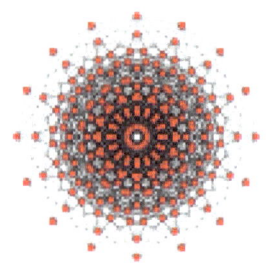

Figure 3-24: 8-cube (Octeract)

Convex regular 9-polytopes are below.

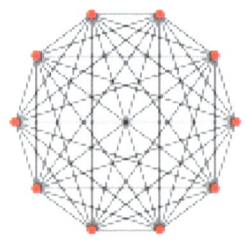

Figure 3-25: 9-simplex (Decayotton)

Figure 3-26: 9-orthoplex (Enneacross)

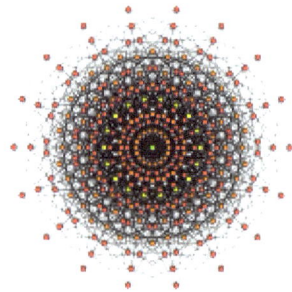

Figure 3-27: 9-cube (Enneract)

Convex regular 10-polytopes are below.

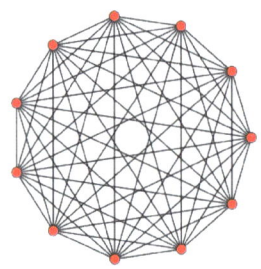 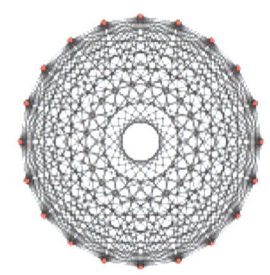 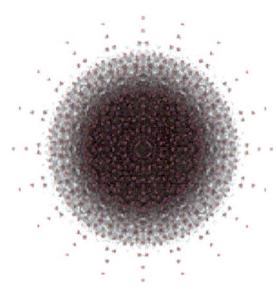

Figure 3-28: 10-simplex (Hendecaxennon)

Figure 3-29: 10-orthoplex (Decacross)

Figure 3-30: 10-cube (Dekeract)

A polytope is a geometric object with flat sides, which exists in any general number of dimensions. A polygon is a polytope in two dimensions. A polyhedron is a polytope in three dimensions. A polychoron is a polytope in four dimensions. It is common to refer to objects of three or more dimensions as polyhedrons, without regard to the actual number of dimensions. Polytopes can increase in dimensions to infinity. An apeirotope is an infinite sided n-polytope called an apeirohedron. While an ordinary polyhedral surface has no border because it folds round to close back on itself, an apeirohedron has no border because its surface is unbounded. It appears to be composed of two matching planes of tiles, infinite in both horizontal and vertical directions.

Polytope is the generic term of a geometric shape that has no curves. A polytope must have flat faces which intersect at straight edges. The straight edges must intersect at vertices. Polytopes must completely enclose an inner region. A polytope can be described as a sequence of increasingly complex objects of point, line segment, polygon, polyhedron, polychoron, and continuing on to apeirohedron.

3-5 Polytope Decomposition

A polytope can be decomposed in to simplex parts. A polytope is a union of simplicies that intersect at vertices, edges, and faces. A point is a 0-simplex, a line segment is a 1-simplex, a triangle is a 2-simplex, a tetrahedron is a 3-simplex, and so forth. Each n-simplex can be reduced or decomposed into a smaller level of simplex. The graph below shows simplexes, from 1 through 10.

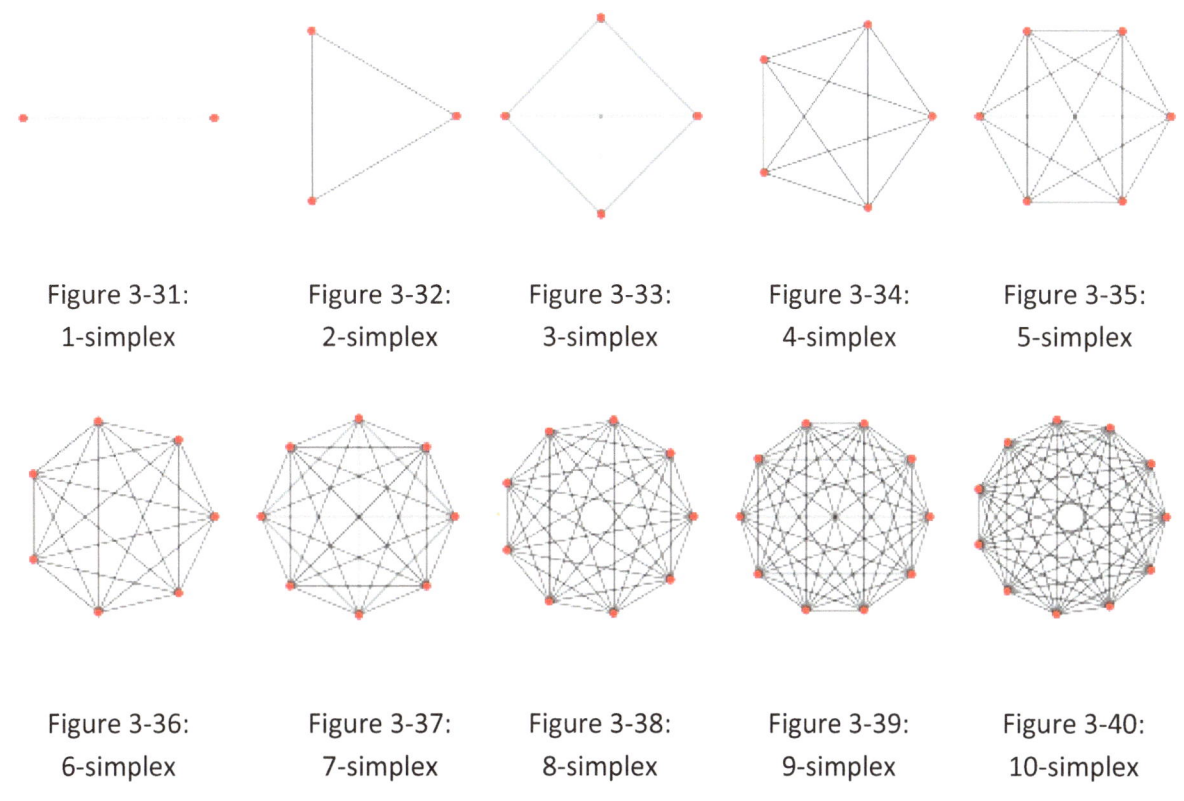

Figure 3-31: 1-simplex
Figure 3-32: 2-simplex
Figure 3-33: 3-simplex
Figure 3-34: 4-simplex
Figure 3-35: 5-simplex

Figure 3-36: 6-simplex
Figure 3-37: 7-simplex
Figure 3-38: 8-simplex
Figure 3-39: 9-simplex
Figure 3-40: 10-simplex

The creation of a polytope starts with a 1-simplex, or line segment. A line segment can be attached to another line segment at the end points or vertices. If two line segments meet at each vertex, a topological curve or polygonal curve is formed. This is also called a polygonal chain because the line segments are attached like a links in a chain, end to end. A series of line segments can be attached by their end points. A curve that does not self-intersect is called a simple curve. When the last line segment is attached to the first line segment, the curve is closed. A simple closed polygonal curve is called a polygon.

In the figures below, line segments form a polygonal curve used to create a polygon.

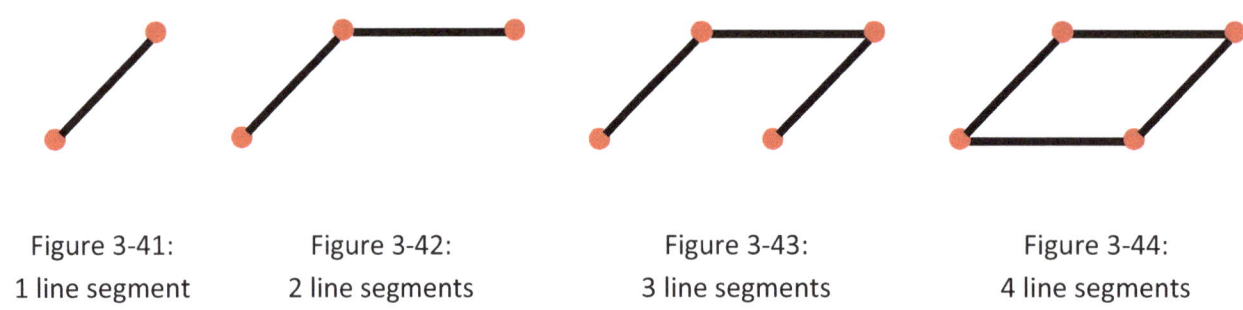

Figure 3-41:
1 line segment

Figure 3-42:
2 line segments

Figure 3-43:
3 line segments

Figure 3-44:
4 line segments

A polygon divides a plane into two areas, the interior of the polygon and the exterior of the polygon. The polygon line, or boundary, and the interior of the polygon are considered to be one object, the polygon. The process can be repeated to create high level polytopes. Polygons are connected along edges or faces to make a polyhedral surface. When the surface is open it is called a skewed polygon and when the surface is closed it is called a polyhedron. Polyhedra can be combined to create polychora. A similar manner can be used to create high levels of polytopes.

- ## 3-6 Summary

A polytope is a geometric object defined as a finite region of n-dimensional space, where *n* is an arbitrary number, enclosed by a finite number of hyperplanes. It is a multidimensional solid with flat sides. A hyperplane is a generalization of the concept of a plane. In zero-dimensional space, the only possible object, or polytope, is a point. In one-dimensional space, the only possible polytopes are points and line segments. In two-dimensional space, there are points, line segments, and polygons. In three-dimensional space, a polyhedron is bounded by polygons, which in turn are bounded by line segments, which in turn are bounded by points. In four-dimensional space, a polychoron is bound by cells (polyhedra).

When referring to an *n*-dimensional generalization, the term *n*-polytope is used. A polytope is a geometric object with flat sides, which exists in any general number of dimensions. A polygon is a polytope in two dimensions. A polyhedron is a polytope in three dimensions. A polychoron is a polytope in four dimensions.

Regular convex polytopes are classified primarily according to their dimensionality. Three special classes of regular polytope exist in every dimensionality: (1) simplex, (2) orthoplex, and (3) hypercube.

Polytope is the generic term of a geometric shape that has no curves. A polytope must have flat faces which intersect at straight edges. The straight edges must intersect at vertices. Polytopes must completely enclose an inner region. A polytope can be described as a sequence of increasingly complex objects of point, line segment, polygon, polyhedron, polychoron, and continuing on to apeirohedron.

A polytope can be decomposed in to simplex parts. A polytope is a union of simplicies that intersect at vertices, edges, and faces. A point is a 0-simplex, a line segment is a 1-simplex, a triangle is a 2-simplex, a tetrahedron is a 3-simplex, and so forth. Each n-simplex can be reduced or decomposed into a smaller level of simplex.

The creation of a polytope starts with a 1-simplex, or line segment. A line segment can be attached to another line segment at the end points or vertices. If two line segments meet at each vertex, a topological curve or polygonal curve is formed. A simple closed polygonal curve is called a polygon.

CHAPTER 3

Chapter Test

Grading Scale: One point for each correct answer.

Excellent = 35-38, Good = 31-34, Average = 27-30, Fair = 23-26, Poor = 0-22

3-1 Introduction

Match definitions and terms.

    A = Polytope   B = Hyperplane   C = Point   D = Line   E = Plane

1. A hyperplane in one-dimensional space. ____
2. A hyperplane in three-dimensional space. ____
3. A multidimensional solid with flat sides. ____
4. A hyperplane in two-dimensional space. ____
5. A generalization of the concept of a plane. ____

Match definitions and terms.

    F = 0-dimension Object   G = 1-dimension Object   H = 2-dimension Object
    I = 3-dimension Object   J = 4-dimension Object

6. Line Segment ____
7. Polygon ____
8. Polychoron ____
9. Polyhedron ____
10. Point ____

3-2 Types of Polytopes

Match definitions and terms.

    A = Monad   B = Polytelon   C = Polygon   D = Polyhedron   E = Polychoron

1. Point ____
2. Pentachoron, Hexadecachoron, Tesseract ____
3. Tetrahedron, Cube, Octahedron ____
4. Triangle, Square, Pentagon ____
5. Line Segment ____

3-3 Regular Convex Polytopes

Match definitions and terms.

    A = Simplex   B = Orthoplex   C = Hypercube

1. Vertices are in pairs orthogonal or perpendicular. _____
2. A generalization of a triangle to an arbitrary dimension. _____
3. Sides are in parallel pairs orthogonal or perpendicular to each other. _____
4. The simplest possible shape for a dimension. _____
5. Cube, Tesseract, Penteract _____
6. Octahedron, Tessacross, Pentacross _____

3-4 Polytopes – Simplex, Orthoplex, and Hypercube

Match definitions and terms.

    A = Pentagon   B = Tetrahedron   C = Hexadecachoron
    D = Hexateron   E = Hexeract   F = Heptacross   G = Octeract
    H = Decayotton   I = Decacross   J = Apeirohedron

1. 6-polytope _____
2. 4-polytope _____
3. Infinite sided n-polytope _____
4. 2-polytope _____
5. 10-polytope _____
6. 7-polytope _____
7. 3-polytope _____
8. 9-polytope _____
9. 8-polytope _____
10. 5-polytope _____

3-5 Polytope Decomposition

Match definitions and terms.

    A = Polygonal Curve   B = Simple Curve   C = Closed Curve
    D = Polyhedral Surface   E = Skewed Polygon   F = Polyhedron
    G = Polychoron

1. The last line segment is attached to the first line segment. _____
2. A closed surface. _____
3. Two line segments meet at each vertex. _____
4. Polyhedra combined together. _____
5. Polygons connected along edges or faces. _____
6. A curve that does not self-intersect. _____
7. An open surface. _____

CHAPTER 4

Polygons

4-1 Introduction

A <mark>polygon</mark> is a two-dimensional polytope or a simple closed polygonal curve on a plane. It is created by attaching a series of straight line segments connected by their end points or vertices. When the last line segment is attached to the first line segment, the curve is closed. The resulting object, called a polygon, has straight sides. The word *polygon* comes from the Greek word πολύς (*polús*) "much", "many" and γωνία (gōnía) "corner" or "angle". Today a polygon is more usually understood in terms of sides.

A polygon can be defined simply as a closed plane figure bounded by three or more line segments. A polygon divides a plane into two areas, the interior of the polygon and the exterior of the polygon. The polygon line, or boundary, and the interior of the polygon are considered to be one object, the polygon.

4-2 Polygon Classification

Polygons can be classified according to their properties. In the following figure, the types of polygons are shown.

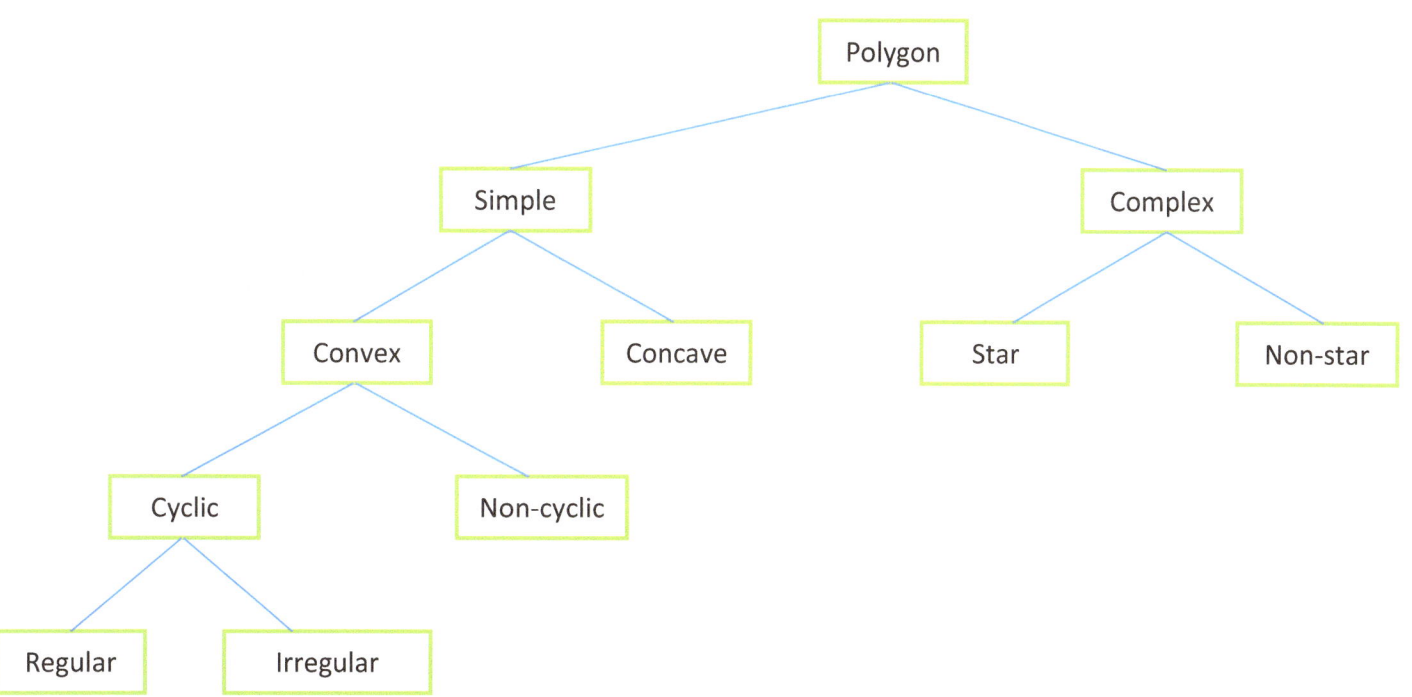

Figure 4-1: Polygon Classification

The classification of polygons starts with determining whether the shape is simple or complex. A simple polygon has a single, non-intersecting boundary. A complex polygon has a self-intersecting boundary. In the figures below, a simple pentagon and a complex pentagon are shown. While both pentagons have five sides, the line segments of the complex pentagon cross over each other.

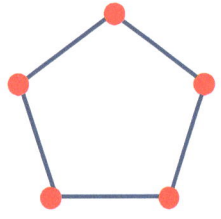
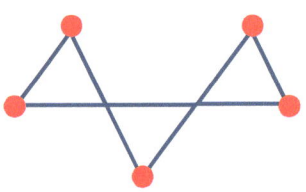

Figure 4-2:
Simple Pentagon Polygon

Figure 4-3:
Complex Pentagon Polygon

Complex polygons can be divided into star shape or non-star shape. In the figures below, two star shape complex polygons and two non-star shape complex polygons are shown.

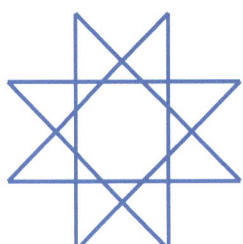
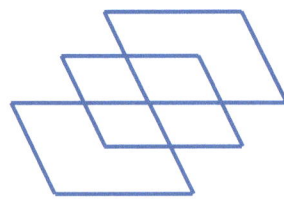

Figure 4-4:
Star Pentagon
(Pentagram)

Figure 4-5:
Star Octagon
(Octagram)

Figure 4-6:
Non-star Polygon – 1

Figure 4-7:
Non-star Polygon – 2

Simple polygons can be divided into convex shape and concave shape. A convex shape has no internal angles greater than 180 degrees. A concave shape has at least one interior angle that exceeds 180 degrees. Another way to determine if a shape is convex or concave is to draw a straight line through the polygon. If it crosses at most two sides, then the polygon is convex. If it crosses more than two sides, the polygon is concave. In the following figures, the two ways of checking for convex or concave polygons are shown.

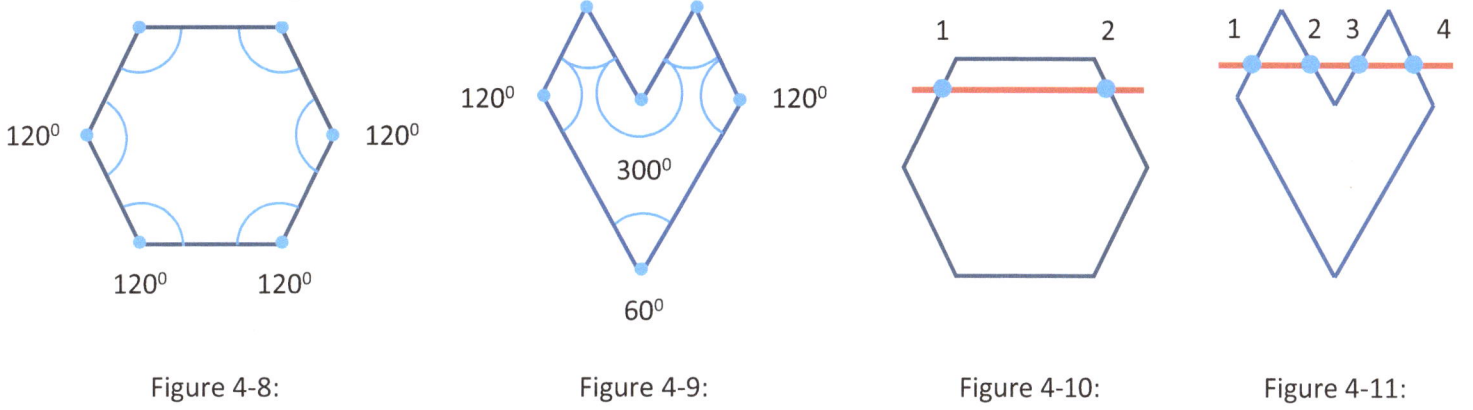

Figure 4-8:
Convex Polygon – 1
No angles greater than 180⁰

Figure 4-9:
Concave Polygon – 1
Angle greater than 180⁰

Figure 4-10:
Convex Polygon – 2
Line crosses two sides

Figure 4-11:
Concave Polygon – 2
Line crosses four sides

Convex polygons can be further divided into cyclic shapes and non-cyclic shapes. If all of the vertices lie on a single circle, the polygon is cyclic. If the vertices do not lie on a single circle, the polygon is non-cyclic. In the following figures, cyclic and non-cyclic polygons are shown.

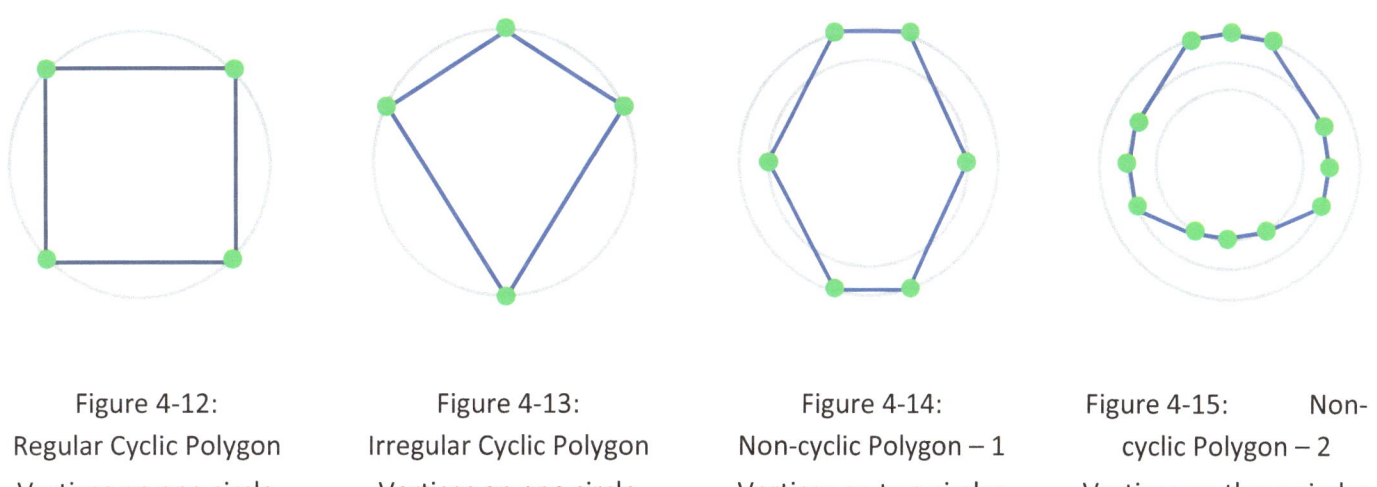

Figure 4-12:
Regular Cyclic Polygon
Vertices on one circle

Figure 4-13:
Irregular Cyclic Polygon
Vertices on one circle

Figure 4-14:
Non-cyclic Polygon – 1
Vertices on two circles

Figure 4-15: Non-cyclic Polygon – 2
Vertices on three circles

Regular polygons have all sides of equal length (equilateral) and all angles are congruent (equiangular). A regular polygon is also called an isogon because it is equilateral and equiangular. Irregular polygons do not have equal side lengths and congruent angles. Because there are many possible types of polygons, most formulas for measuring polygons are based on using regular polygons.

4-3 Polygon Parts

Polygons have several parts which need to be identified. A side is a line segment that forms the edge of the polygon. For example, a triangle has three sides. A vertex is the intersection of two sides at a point. For example, a square has four vertices. An interior angle is formed by two adjacent sides inside the polygon. For example, the interior angle of a regular octagon is 135 degrees. An outside angle is formed by two adjacent sides outside of the polygon. The outside angle is the explementary angle to the interior angle. For example, the outside angle of a regular octagon is 225 degrees. An exterior angle is an angle formed on the outside of a polygon between a side and the extended adjacent side. The exterior angle is the supplementary angle to the interior angle. For example, the exterior angle of a regular octagon is 45 degrees. A diagonal is a line connecting two vertices that are not a side. Diagonals can be drawn to connect each of the vertices. When diagonals are drawn from only one vertex, the polygon is divided into triangles. In the first figure below, the polygon parts are shown. In the second figure below, the diagonals of an octagon are shown. In the third figure below, the triangles of an octagon are shown.

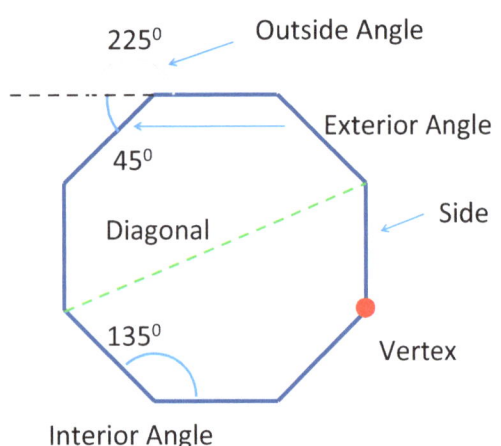
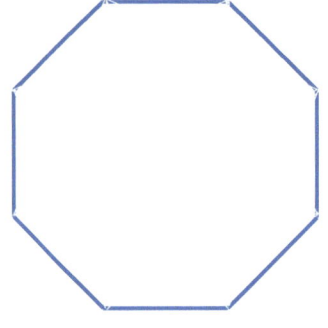
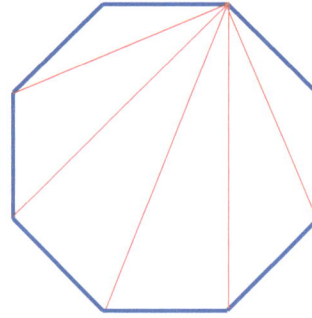

Figure 4-16:
Polygon Parts
Regular Octagon

Figure 4-17:
Diagonals of Octagon
Diagonals = 20

Figure 4-18:
Triangles of Octagon
Triangles = 6

4-4 Polygon Formulas

The interior angle degrees, exterior degrees, number of diagonals, and number of triangles can be calculated for a regular polygon using formulas. In the following table, formulas for regular polygons are shown.

Polygon Formulas (N = number of sides)	N = 3	N = 4	N = 5	N = 6	N = 7	N = 8
Sum of interior angles: (N-2) x 180	180	360	540	720	900	1080
Interior angle: [(N-2) x 180] / N	60	90	108	120	128.5	135
Outside angle: 360 – interior angle	300	270	252	240	231.5	225
Exterior angle: 180 – interior angle	120	90	72	60	51.5	45
Number of diagonals: (1/2) x N(N-3)	0	2	5	9	14	20
Number of triangles: N-2	1	2	3	4	5	6

Table 4-1: Polygon Formulas

4-5 Polygon Names

The names of polygons can take several different formats. The names can become quite long as the number of sides or angles increase. In general, most polygons with greater than twelve sides are referred to by sides in an n-gon format rather than by name, unless the name is easily pronounceable or spelled. The n-gon format uses the number of sides, as a number, followed by a hyphen and the -gon base. For example, a triskaidecagon is more commonly called a 13-gon. For three and four sided polygons, which are commonly called triangle and quadrilateral, the -gon base is not used in the common name. This is inconsistent with the names of other polygons, but is used based on traditional naming. In the table below, polygon names are shown.

Number of sides or angles	Greek prefix + English base	Greek prefix + Greek base	Greek prefix + Latin base	n-gon format	Common or preferred name
3	triangle	trigon	trilateral	3-gon	triangle
4	quadrangle	tetragon	quadrilateral	4-gon	quadrilateral
5	pentangle	pentagon	pentalateral	5-gon	pentagon
6	hexangle	hexagon	hexalateral	6-gon	hexagon

7	heptangle	heptagon	heptalateral	7-gon	heptagon
8	octangle	octagon	octalateral	8-gon	octagon
9	enneangle	enneagon	ennealateral	9-gon	enneagon
10	decangle	decagon	decalateral	10-gon	decagon
11	hendecangle	hendecagon	hendecalateral	11-gon	hendecagon
12	dodecangle	dodecagon	dodecalateral	12-gon	dodecagon
13	tridecangle	triskaidecagon	tridecalateral	13-gon	13-gon
14	tetradecangle	tetradecagon	tetradecalateral	14-gon	14-gon
15	pentadecangle	pentadecagon	pentadecalateral	15-gon	15-gon
16	hexadecangle	hexadecagon	hexadecalateral	16-gon	16-gon
17	heptadecangle	heptadecagon	heptadecalateral	17-gon	17-gon
18	octadecangle	octadecagon	octadecalateral	18-gon	18-gon
19	enneadecangle	enneadecagon	enneadecalateral	19-gon	19-gon
20	isosangle	isosagon	isosalateral	20-gon	isosagon
30	triacontrangle	triacontagon	tricontalateral	30-gon	triacontagon
40	tetracontrangle	tetracontagon	tetracontalateral	40-gon	tetracontagon
50	pentacontrangle	pentacontagon	pentacontalateral	50-gon	pentacontagon
60	hextracontrangle	hextacontagon	hextracontalateral	60-gon	hextacontagon
70	heptacontrangle	heptacontagon	heptacontalateral	70-gon	heptacontagon
80	octacontrangle	octacontagon	octacontalateral	80-gon	octacontagon
90	enneacontrangle	enneacontagon	enneacontalateral	90-gon	enneacontagon
100	hectangle	hectagon	hectalateral	100-gon	hectagon
1,000	chiliangle	chiliagon	chilialateral	1000-gon	1000-gon
10,000	myriangle	myriagon	myrialateral	10000-gon	10000-gon

Table 4-2: Polygon Names

Polygon names can be constructed for polygons with more than 20 and less than 100 sides by combining prefixes with the –gon base. In the following table, the method for constructing additional polygon names is shown.

Tens	Name	And	Ones	Name	Base
			1	hena-	gon
20	isosi-	kai-	2	di-	gon
30	triconta-	kai-	3	tri-	gon
40	tetraconta-	kai-	4	tetra-	gon
50	pentaconta-	kai-	5	penta-	gon
60	hexaconta-	kai-	6	hexa-	gon
70	heptaconta-	kai-	7	hepta-	gon
80	octaconta-	kai-	8	octa-	gon
90	enneaconta-	kai-	9	ennea-	gon

Table 4-3: Polygon Naming Method

Using the above table, polygons with any number of sides can be named. A polygon with 24 sides is called an isosikaitetragon, or a 24-gon. A polygon with 58 sides is called a pentacontakaioctagon, or a 58-gon. A polygon with 79 sides is called a heptacontakaienneagon, or a 79-gon. A polygon with 365 sides is called a triacosiakaihexacontakaipentagon, or a 365-gon. A polygon with an infinite number of sides is called an apeirogon, and it appears as a straight line.

• 4-6 Regular Polygon Properties

The area of a polygon is the measurement of the region enclosed by the polygon or the amount of units within the object. There are several formulas for calculating the area of specific types of regular polygons. The formula for calculating the area for all regular polygons is as follows: area = 1/2 x perimeter x apothem. The perimeter can be calculated by multiplying the number of sides by the length of a side. Area is a capital (upper case) letter a (A) in formulas. Sometimes a capital K is used for area, if a capital letter a (A) is used to label part of a polygon. The perimeter of a polygon, or measure around the outside, is calculated by adding the length of each side of the polygon. Perimeter is a small (lower case) letter p (p) in formulas.

The apothem is the distance from the center of a polygon to the midpoint of a side. It is also the inradius of the incircle. The small letter a (a) for apothem or small letter r (r) for inradius is used in formulas. The radius is the distance from the center of the polygon to any vertex. It is also the circumradius of the circumcircle. Radius is a capital r (R) in formulas. The incircle is a circle inside of a polygon that touches each of the sides. The circumcircle is a circle outside of a polygon that intersects each of the vertices.

The figures below show each of these six properties of regular polygons.

 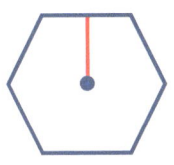 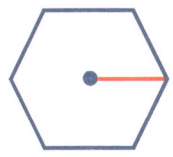 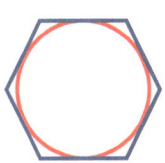 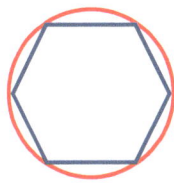

Figure 4-19: Area = 9   Figure 4-20: Perimeter = 10   Figure 4-21: Apothem   Figure 4-22: Radius   Figure 4-23: Incircle   Figure 4-24: Circumcircle

The central angle is at the center of a polygon made by two adjacent radius lines. The degree measure of the central angle can be calculated with the following equation: degrees of central angle = 360 degrees / number of sides. For example, the central angle of a pentagon is calculated as 360 / 5 = 72 degrees. In the figures below, the central angle is shown for four regular polygons.

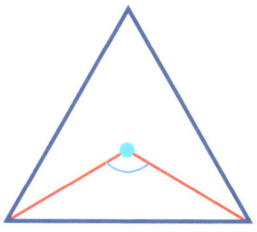
Figure 4-25:
Central angle = 120°

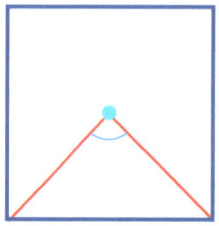
Figure 4-26:
Central angle = 90°

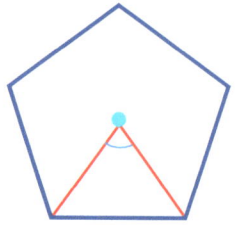
Figure 4-27:
Central angle = 72°

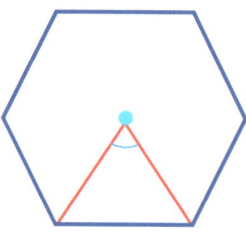
Figure 4-28:
Central angle = 60°

The figures below show common types of regular polygons.

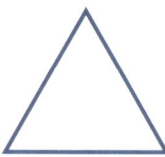
Figure 4-29:
Triangle
Sides = 3

Figure 4-30:
Square
Sides = 4

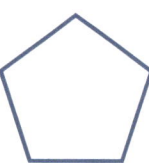
Figure 4-31:
Pentagon
Sides = 5

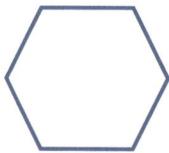
Figure 4-32:
Hexagon
Sides = 6

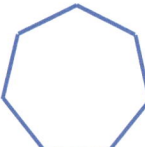
Figure 4-33:
Heptagon
Sides = 7

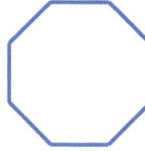
Figure 4-34:
Octagon
Sides = 8

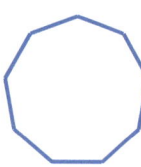
Figure 4-35:
Enneagon
Sides = 9

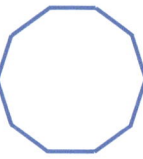
Figure 4-36:
Decagon
Sides = 10

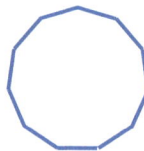
Figure 4-37:
Hendecagon
Sides = 11

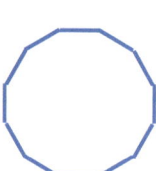
Figure 4-38:
Dodecagon
Sides = 12

• 4-7 Triangle

A triangle is the simplest or most basic polygon. A regular triangle is a polygon with three equal sides and three congruent angles. The interior angles are 60 degrees and the exterior angles are 120 degrees. There are many types of triangles based on the lengths of the sides and the measurements of the angles. A general triangle is a triangle that represents all triangles. This means that descriptions, properties, and formulas that apply to general triangles apply to all triangles, regardless of size or angles. An inscribed triangle contains an incircle. A circumscribed triangle is inside of a circumcircle. In the figures below, a general triangle, inscribed triangle, and circumscribed triangle are shown.

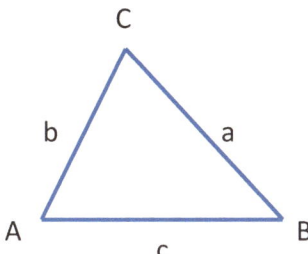

Figure 4-39:
General triangle

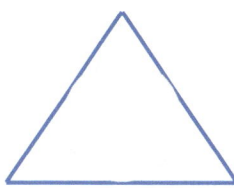

Figure 4-40:
Inscribed triangle

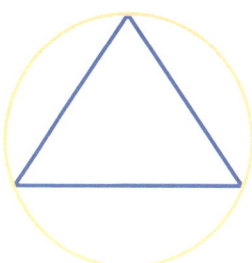

Figure 4-41:
Circumscribed triangle

Each vertex of a triangle is named with a single capital (upper case) letter. The sides are named with a single small (lower case) letter corresponding to the opposite angle. In the first figure above, side b is opposite angle B. The sides can also be labeled using the name of the line segment. Side b can also be called side AC.

The base of a triangle can be any one of the three sides, but is usually drawn at the bottom. The altitude of a triangle is the height. The altitude is perpendicular from the base to the opposite vertex. The altitude line divides a triangle into two opposite facing right triangles. The angle bisector is a transitive line that divides the angle into two equal parts. The median of a triangle is a line from a vertex to the midpoint of the opposite side that divides the base into two equal parts. The altitude, angle bisector, and median are used in triangle area formulas. In most triangles, the altitude, angle bisector, and median lines will be different lines. In regular triangles, each of these three lines is the exact same line. In the figures below, general triangles are shown with altitude, angle bisector, and median lines on side c or side AB.

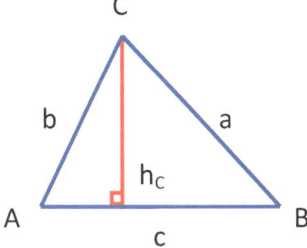

Figure 4-42:
General triangle
Altitude line

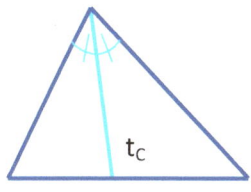

Figure 4-43:
General triangle
Angle bisector line

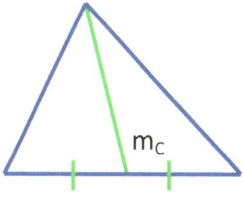

Figure 4-44:
General triangle
Median line

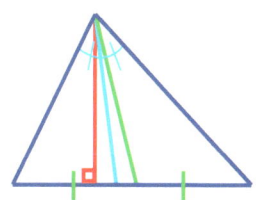

Figure 4-45:
General triangle
All three lines

A regular triangle is both equilateral and equiangular. Equilateral means all sides are equal and equiangular mean all interior angles are equal or congruent. In a regular triangle, all three sides are labeled as side *a* because they are of equal length. The altitude line divides the base into two equal parts. A right triangle has one angle equal to 90 degrees. In a right triangle, the altitude line divides the base into two parts, labeled *m* and *n*, which may not be equal. In the figures below, a regular triangle and a right triangle are shown.

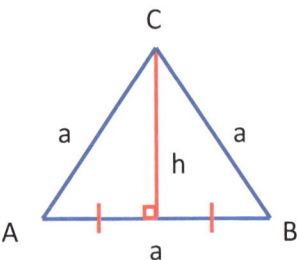

Figure 4-46: Regular triangle

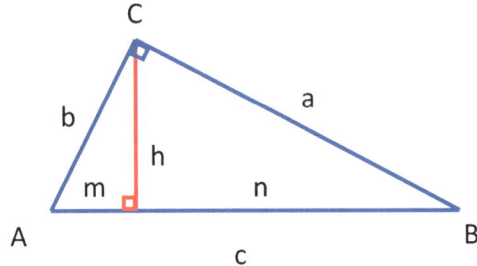

Figure 4-47: Right triangle

• 4-8 Triangle Formulas

The tables below show formulas for calculating triangle measurements.

General Triangle Formulas	
$A + B + C = 180^0$	
$K = \sqrt{s(s-a)(s-b)(s-c)}$ with $s = 1/2\,(a+b+c)$	Heron's Formula
$K = 1/2\,\sqrt{a^2c^2 - (((a^2 + c^2 - b^2)/2)^2)}$	Qin Jiushao's Formula
$r = K/s$	
$R = abc/4K$	
$hc = 2K/c$	
$tc = \sqrt{ab(1 - (c^2/(a+b)^2))}$	
$mc = \sqrt{(a^2/2) + (b^2/2) - (c^2/4)}$	

Regular Triangle Formulas
$A = B = C = 60^0$
$K = 1/4\,a^2\,\sqrt{3}$
$K = 1/2\,ah$
$r = 1/6\,a\,\sqrt{3}$
$R = 1/3\,a\,\sqrt{3}$
$h = 1/2\,a\,\sqrt{3}$

Right Triangle Formulas
A + B = C = 90⁰
$c^2 = a^2 + b^2$
a = √((c + b)(c − b))
K = 1/2 ab
r = ab / (a + b + c)
R = 1/2 c
h = ab / c
m = b^2 /c
n = a^2 / c

Table 4-4: Triangle Formulas

The measurements of a regular triangle can be made using either the general triangle formulas or the regular triangle formulas. The measurements of a right triangle can be made using either the general triangle formulas or the right triangle formulas. The formula that is used is based on the known measurements of a triangle, or which formula is easier to use. In the figures below, a regular triangle and a right triangle are shown. These sample triangles are used for calculations below.

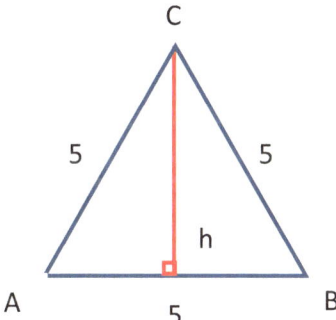

Figure 4-48: Regular triangle – Sample values

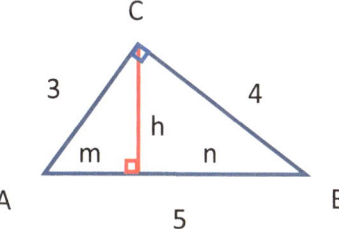

Figure 4-49: Right triangle – Sample values

The area of the regular triangle can be calculated using the general triangle formula as follows.

K = √(s(s − a)(s − b)(s − c)) with s = 1/2 (a + b + c)

K = √(7.5(7.5 − 5)(7.5 − 5)(7.5 − 5)) with s = 1/2 (5 + 5 + 5)

K = √(117.1875) = 10.82531755

The area of a regular triangle can be calculated using the regular triangle formula as follows.

$K = 1/4\ a^2\ \sqrt{3}$

$K = 1/4\ (5^2)\ (1.732050808)$

$K = 1/4\ (43.30127019) = 10.82531755$

The area of the right triangle can be calculated using the general triangle formula as follows.

$K = \sqrt{(s(s-a)(s-b)(s-c))}$ with $s = 1/2\ (a + b + c)$

$K = \sqrt{(6(6-4)(6-3)(6-5))}$ with $s = 1/2\ (4 + 3 + 5)$

$K = \sqrt{(36)} = 6$

The area of the right triangle can be calculated using the right triangle formula as follows.

$K = 1/2\ ab$

$K = 1/2\ (4)(3)$

$K = 1/2\ (12) = 6$

• 4-9 Trigonometry

Trigonometry (from Greek *trigōnon* "triangle" + *metron* "measure") is a branch of mathematics that studies triangle and the relationships between their sides and the angles between these sides. The trigonometric functions comprising trigonometry are the sine (sin x), cosine (cos x), tangent (tan x), cosecant (csc x), secant (sec x), and cotangent (cot x). The inverses of these functions are denoted as $\sin^{-1}$, $\cos^{-1}$, $\tan^{-1}$, $\csc^{-1}$, $\sec^{-1}$, and $\cot^{-1}$. The figures below show a right triangle with trigonometry relationships.

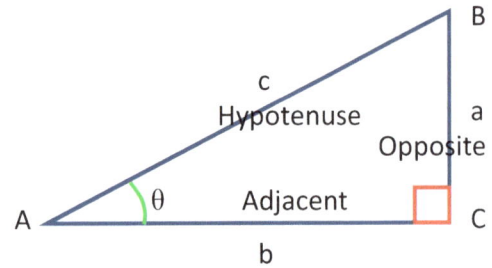

Sin A = a/c	Csc A = c/a
Cos A = b/c	Sec A = c/b
Tan A = a/b	Cot A = b/a

Figure 4-50: Trigonometry Triangle   Figure 4-51: Trigonometric Functions

Discovery Activity.

1. Discover the history of Trigonometry. Important contributions were made by several civilizations over a long period of time. Conduct research to explore the development of trigonomic principles of each civilization.
    a. Make a table show the contributions for the following civilizations: Egyptian (300-100 BCE), Babylonian (300-100 BCE), Greek (300-100 BCE), Hindu (400-700 CE), Chinese (700-1300 CE), Islamic (900-1300 CE), and European (1300-1800 CE).
    b. Make a timeline to show a combined list of the contributions from all of the civilizations in chronological order.
2. Read introductions to trigonometry. They are available in print and online from many sources.
    a. Make a list of the basic topics in trigonometry.
    b. Make a list and summarize the most important trigonomic functions.

4-10 Regular Polygon Areas

In the table below, formulas for calculating the areas of polygons are shown. In the calculations, s = length of side.

Polygon	Sides	Area Formula	Example
Triangle	3	$0.433012702 \times s^2$	$s = 5, A = 0.433012702 \times 5^2 = 10.82531755$
Square	4	$1 \times s^2$	$s = 4, A = 1 \times 4^2 = 16$
Pentagon	5	$1.720477401 \times s^2$	$s = 3, A = 1.720477401 \times 3^2 = 15.48429661$
Hexagon	6	$2.598076211 \times s^2$	$s = 2, A = 2.598076211 \times 2^2 = 10.39230485$
Heptagon	7	$3.633912444 \times s^2$	$s = 5, A = 3.633912444 \times 5^2 = 90.8478111$
Octagon	8	$4.828427125 \times s^2$	$s = 4, A = 4.828427125 \times 4^2 = 77.254834$
Enneagon	9	$6.181824194 \times s^2$	$s = 3, A = 6.181824194 \times 3^2 = 55.63641774$
Decagon	10	$7.694208843 \times s^2$	$s = 2, A = 7.694208843 \times 2^2 = 30.77683537$
Hendecagon	11	$9.365639907 \times s^2$	$s = 5, A = 9.365639907 \times 5^2 = 234.1409977$
Dodecagon	12	$11.19615242 \times s^2$	$s = 4, A = 11.19615242 \times 4^2 = 179.1384388$

Table 4-5: Polygon Calculations

4-11 Drawing Regular Polygons

It is often useful to draw a regular polygon for visualizing the solution to a problem. The only tools needed are a protractor and a pencil. The following instructions will allow for the construction of a regular polygon with any number of sides and of any radius.

1. Choose the number of sides of the polygon and calculate the central angle.

    Central angle = $2\pi/n * 180/\pi = 360/n$, with n = number of sides

    For example, the central angle of a hexagon is 360/6 degrees or 60 degrees.

2. Choose the center of the polygon and length of the radius. The radius is the distance from the center to a vertex.
3. Place the center point of the protractor at the location of center of the polygon. Draw a dot for the center of the polygon.
4. Draw a dot at the central angle and each multiple of the central angle around a circle.

    For example, mark the following degree points: 60, 120, 180, 240, 300, and 360.

5. Draw a line from the center of the polygon to each marked degree point.
6. Measure along each line the length of the radius and draw a point.
7. Connect the radius points with straight lines. The polygon is drawn.
8. Erase any unneeded lines or dots.

The figures below show the drawing of a hexagon using the instructions above.

Step 3: Draw center of polygon.

Step 4: Draw each degree point.

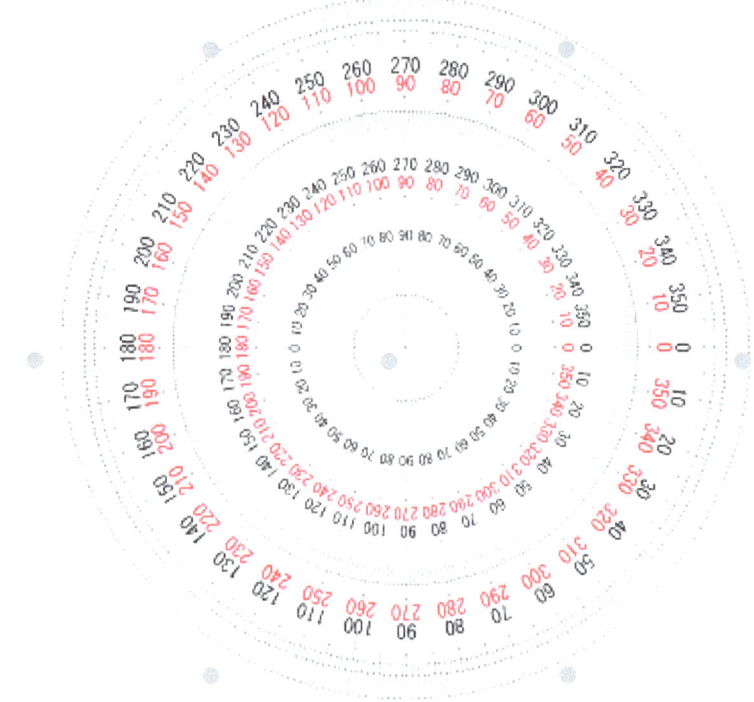

Figure 4-52: Drawing Steps 3 and 4

Step 5: Draw line from center to each degree point.

Step 6: Draw radius length point on each line.

Step 7: Draw line connecting each radius point.

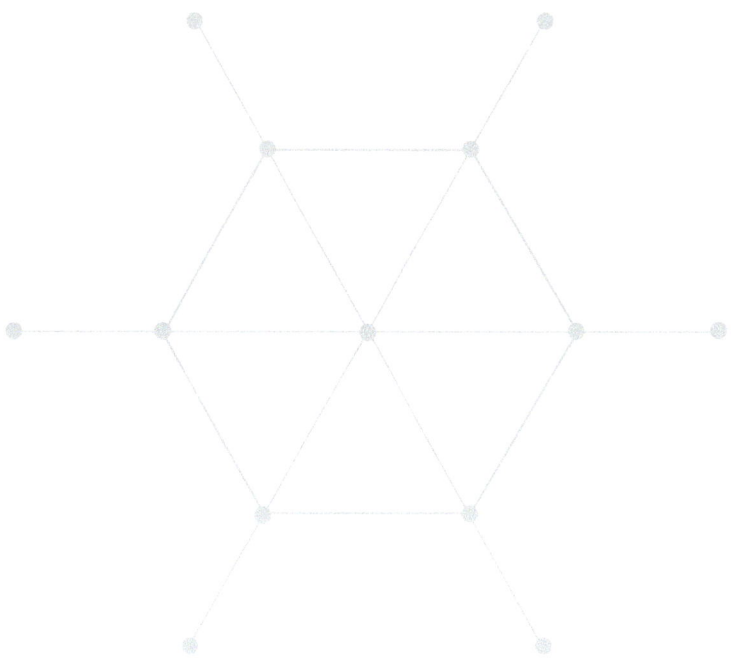

Figure 4-53: Drawing Steps 5, 6, and 7

4-12 Summary

A polygon is a two-dimensional polytope or a simple closed polygonal curve on a plane. It is created by attaching a series of straight line segments connected by their end points or vertices. When the last line segment is attached to the first line segment, the curve is closed. The resulting object, called a polygon, has straight sides. A polygon can be defined simply as a closed plane figure bounded by three or more line segments. A polygon divides a plane into two areas, the interior of the polygon and the exterior of the polygon.

Polygons can be classified according to their properties. The classification of polynomials starts with determining whether the shape is simple or complex. A simple polygon has a single, non-intersecting boundary. A complex polygon has a self-intersecting boundary. Complex polygons can be divided into star shape or non-star shape. Simple polygons can be divided into convex shape and concave shape. A convex shape has no internal angles greater than 180 degrees. A concave shape has at least one interior angle that exceeds 180 degrees. Convex polygons can be further divided into cyclic shapes and non-cyclic shapes. If all of the vertices lie on a single circle, the polygon is cyclic. If the vertices do not lie on a single circle, the polygon is non-cyclic.

Regular polygons have all sides of equal length (equilateral) and all angles are congruent (equiangular). A regular polygon is also called an isogon because it is equilateral and equiangular. Irregular polygons do not have equal side lengths and congruent angles.

Polygons have several parts which need to be identified. A side is a line segment that forms the edge of the polygon. A vertex is the intersection of two sides at a point. An interior angle is formed by two adjacent sides inside the polygon. An outside angle is formed by two adjacent sides outside of the polygon. The outside angle is the explementary angle to the interior angle. An exterior angle is an angle formed on the outside of a polygon between a side and the extended adjacent side. The exterior angle is the supplementary angle to the interior angle. A diagonal is a line connecting two vertices that are not a side. Diagonals can be drawn to connect each of the vertices. When diagonals are drawn from only one vertex, the polygon is divided into triangles. The interior angle degrees, exterior degrees, number of diagonals, and number of triangles can be calculated for a regular polygon using formulas.

The names of polygons can take several different formats. The names can become quite long as the number of sides or angles increase. In general, most polygons with greater than twelve sides are referred to by sides in an n-gon format rather than by name, unless the name is easily pronounceable or spelled.

The area of a polygon is the measurement of the region enclosed by the polygon or the amount of units within the object. There are several formulas for calculating the area of specific types of regular polygons. The perimeter is the measure around the outside of a polygon. The apothem is the distance from the center of a polygon to the midpoint of a side. It is also the inradius of the incircle. The radius is the distance from the center of the polygon to any vertex. It is also the circumradius of the circumcircle. The incircle is a circle inside of a polygon that touches each of the sides. The circumcircle is a circle outside of a polygon that intersects each of the vertices. The central angle is at the center of a polygon made by two adjacent radius lines.

A triangle is the simplest or most basic polygon. A regular triangle is a polygon with three equal sides and three congruent angles. The interior angles are 60 degrees and the exterior angles are 120 degrees. There are many types of triangles based on the lengths of the sides and the measurements of the angles. A general triangle is a triangle that represents all triangles. This means that descriptions, properties, and formulas that apply to general triangles apply to all triangles, regardless of size or angles. An inscribed triangle contains an incircle. A circumscribed triangle is inside of a circumcircle.

The base of a triangle can be any one of the three sides, but is usually drawn at the bottom. The altitude of a triangle is the height. The altitude is perpendicular from the base to the opposite vertex. The altitude line divides a triangle into two opposite facing right triangles. The angle bisector is a transitive line that divides the angle into two equal parts. The median of a triangle is a line from a vertex to the midpoint of the opposite side that divides the base into two equal parts. The altitude, angle bisector, and median are used in triangle area formulas. In most triangles, the altitude, angle bisector, and median lines will be different lines. In regular triangles, each of these three lines is the exact same line.

A regular triangle is both equilateral and equiangular. Equilateral means all sides are equal and equiangular mean all interior angles are equal or congruent. In a regular triangle, all three sides are labeled as side *a* because they are of equal length. The altitude line divides the base into two equal parts. A right triangle has one angle equal to 90 degrees. In a right triangle, the altitude line divides the base into two parts, labeled *m* and *n*, which may not be equal.

The measurements of a regular triangle can be made using either the general triangle formulas or the regular triangle formulas. The measurements of a right triangle can be made using either the general triangle formulas or the right triangle formulas. The formula that is used is based on the known measurements of a triangle, or which formula is easier to use. Trigonometry is a branch of mathematics that studies triangle and the relationships between their sides and the angles between these sides.

It is often useful to draw a regular polygon for visualizing the solution to a problem. The only tools needed are a protractor and a pencil. Instructions allow for the construction of a regular polygon with any number of sides and of any radius.

CHAPTER 4 — Chapter Test

Grading Scale: One point for each correct answer.

Excellent = 79-87, Good = 70-78, Average = 61-69, Fair = 53-60, Poor = 0-52

4-2 Polygon Classification

Match definitions and terms.

> A = Simple Polygon   B = Complex Polygon   C = Convex Polygon
> D = Concave Polygon   E = Cyclic Polygon   F = Regular Polygon

1. All sides of equal length and all angles are congruent. _____
2. At least one interior angle that exceeds 180 degrees. _____
3. A single, non-intersecting boundary. _____
4. No internal angles greater than 180 degrees. _____
5. All of the vertices lie on a single circle. _____
6. A self-intersecting boundary. _____

4-3 Polygon Parts

Calculate the exterior and outside angles of a regular polygon.

1. Interior angle = 60^0 _____ degrees   _____ degrees
2. Interior angle = 90^0 _____ degrees   _____ degrees
3. Interior angle = 108^0 _____ degrees   _____ degrees
4. Interior angle = 120^0 _____ degrees   _____ degrees

4-4 Polygon Formulas

Calculate the sum of the interior angles, number of diagonals, and number of triangles.

1. Number of sides = 9   _____ degrees   _____   _____
2. Number of sides = 10   _____ degrees   _____   _____
3. Number of sides = 11   _____ degrees   _____   _____
4. Number of sides = 12   _____ degrees   _____   _____

Calculate the interior angle, outside angle, and exterior angle.

5. Number of sides = 9   _____ degrees   _____ degrees   _____ degrees

6. Number of sides = 10   _____ degrees   _____ degrees   _____ degrees
7. Number of sides = 11   _____ degrees   _____ degrees   _____ degrees
8. Number of sides = 12   _____ degrees   _____ degrees   _____ degrees

4-5 Polygon Names

Construct the polygon names.

1. Number of sides = 36   _____
2. Number of sides = 48   _____
3. Number of sides = 72   _____
4. Number of sides = 91   _____

4-6 Regular Polygon Properties

Calculate the area of a regular polygon.

1. Perimeter = 20 units, apothem = 5 units   _____ square units
2. Perimeter = 10 units, apothem = 3 units   _____ square units
3. Perimeter = 14 units, apothem = 1.5 units   _____ square units
4. Perimeter = 7.5 units, apothem = 2 units   _____ square units

Calculate the central angle of a regular polygon.

5. Pentagon   _____ degrees
6. Hexagon   _____ degrees
7. Heptagon   _____ degrees
8. Octagon   _____ degrees

4-7 Triangle

Match definitions and terms.

A = Regular Triangle   B = General Triangle   C = Inscribed Triangle
D = Circumscribed Triangle   E = Altitude   F = Angle Bisector
G = Median   H = Equilateral   I = Equiangular

1. All interior angles are congruent. _____
2. A transitive line that divides the angle into two equal parts. _____
3. A triangle that represents all triangles. _____
4. Triangle is inside of a circumcircle. _____
5. A line from a vertex to the midpoint of the opposite side. _____
6. A polygon with three equal sides and three congruent angles. _____
7. All sides are equal. _____
8. Triangle contains an incircle. _____
9. A line perpendicular from the base to the opposite vertex. _____

4-8 Triangle Formulas

Calculate the area (K), inradius (r), and circumradius (R) for a triangle.

1. General triangle with a = 5, b = 7, c = 9.

   _____ square units   _____ units   _____ units

2. Regular triangle with a = 8, b = 8, c = 8.

   _____ square units   _____ units   _____ units

3. Right triangle with a = 6, b = 8, c = 10.

   _____ square units   _____ units   _____ units

4-10 Regular Polygon Areas

Calculate the area of a regular polygon. (Round to the nearest thousandth.)

1. Dodecagon with side = 6.   _____ square units
2. Square with side = 3.   _____ square units
3. Hendecagon with side = 3.   _____ square units
4. Pentagon with side = 4.   _____ square units
5. Enneagon with side = 4.   _____ square units
6. Hexagon with side = 5.   _____ square units
7. Octagon with side = 5.   _____ square units
8. Heptagon with side = 2.   _____ square units
9. Triangle with side = 6.   _____ square units

4-11 Drawing Regular Polygons

Calculate the central angle for polygons. The formula is as follows: $2\pi/n * 180/\pi = 360/n$, with n = number of sides.

1. Triangle   ____ degrees
2. Square   ____ degrees
3. Pentagon   ____ degrees
4. Hexagon   ____ degrees
5. Heptagon   ____ degrees
6. Octagon   ____ degrees
7. Enneagon   ____ degrees
8. Decagon   ____ degrees
9. Hendecagon   ____ degrees
10. Dodecagon   ____ degrees

CHAPTER 5: Triangles and Quadrilaterals

Section 1 – Triangles

5-1 Introduction

A triangle is a polygon with three sides and three angles. There are many types of triangles based on the lengths of the sides and the measurements of the angles. A triangle with vertices *A*, *B*, and *C* is denoted $\triangle ABC$. All triangles are convex and bicentric. Bicentric means a polygon possesses both an incircle and a circumcircle. The sum of all angles in any triangle is 180 degrees. The largest angle is at least 60 degree. The angle opposite the longest side is the largest angle. The length of any given side of a triangle is less than the sum of the lengths of the remaining two sides. The difference between the lengths of any two sides of a triangle is less than the length of the third side. In the first figure below, a triangle is shown with marked sides and vertices. In the second figure below, a triangle is shown with a marked incenter and a circumcenter.

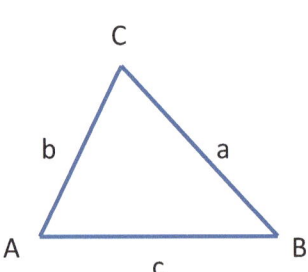

Figure 5-1: General triangle

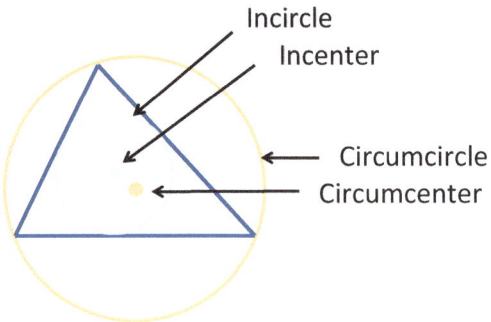

Figure 5-2: Triangle incenter and circumcenter

5-2 Triangle Types

Triangles can be divided into types based on the length of sides or the measurement of the angles. A triangle can be a combination of types. Based on the length of the sides, a triangle can be equilateral, isosceles, or scalene. An equilateral triangle has three sides of equal length and is a regular polygon. Every equilateral triangle is also an equiangular triangle, with all three angles measuring 60 degrees.

An isosceles triangle has at least two side of equal length. The angles opposite the equal sides are congruent. These angles are usually referred to as base angles. In an isosceles triangle, the median, bisectrix, and height lines of the vertex between the equal sides all coincide. This median/bisectrix/height line divides the isosceles triangle into two congruent right triangles. If a median coincides with a bisectrix, or if a median coincides with a height, or if a bisectrix coincides with a height, then the triangle is isosceles and the adjacent sides are equal. The medians of the equal sides of an isosceles triangle are equal. The heights and bisectors crossing the equal sides are equal also. An equilateral triangle is a special case of an isosceles triangle having not just two, but all three sides equal.

A scalene triangle has all three sides of unequal length. The three angles are all different in measurement. If a triangle has no special properties of equal side lengths or congruent angles, then the triangle is scalene. Interesting fact of equilateral triangles: The sum of the distances from any point in the interior of an equilateral triangle to all three sides is always equal to the height of the equilateral triangle. In the figures and box below, the proof of this equality is shown.

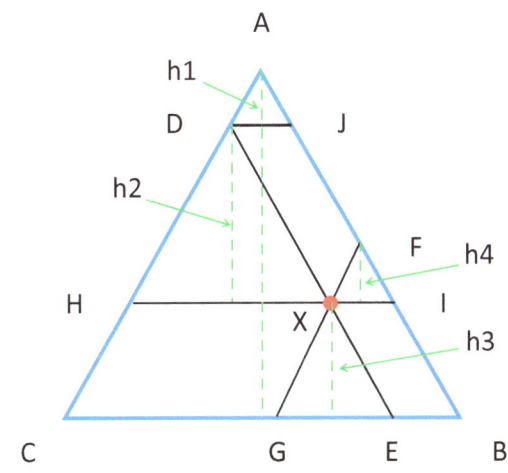

Figure 5-3: Triangle ABC – Methods 1 and 2

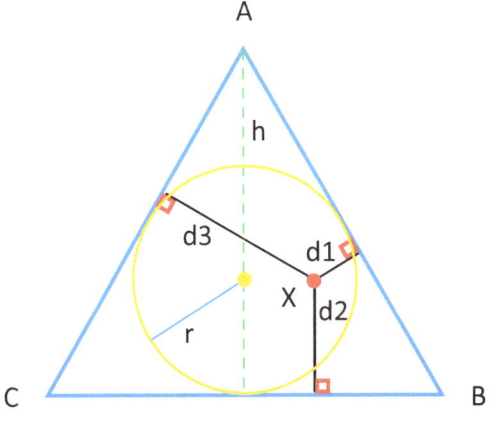

Figure 5-4: Triangle ABC – Method 3

Triangle ABC is an equilateral triangle. Select point X inside of the triangle. Draw line DE parallel to AB, FG parallel to AC, and HI parallel to BC, with each line going through point X. Triangles DHX, EGX, and FIX are similar to ABC, and are equilateral. To prove the sum of the distances from point X to all three sides is equal to the height of triangle ABC, there are three methods.

Method 1: Denote the height of triangle ABC as h1, DHX as h2, EGX as h3, and FIX as h4. Equivalent ratios of triangle ABC are $h2/h1 = DH/AC$, $h3/h1 = GE/BC$, and $h4/h1 = FI/AB$. These equalities can be added. Because BEXI and CHXG are parallelograms, $IX = BE$ and $HX = CG$. Since triangles ABC, DHX, EGX, and FIX are equilateral, the equality becomes $(h2+h3+h4)/h1$.

Method 2: Draw line DJ parallel to CB. Triangles ADJ and FIX are congruent. The combined heights of triangles EGX, DHX, and ADJ exactly equal the height of triangle ABC.

Method 3: In a regular or equilateral triangle, formulas can be used for the height (h) and inradius (r). With the length of a side represented by a lower case a (a), $h = 1/2\ a\ \sqrt{3}$ and $r = 1/6\ a\ \sqrt{3}$. As a result $h = 3r$. The distances (d1, d2, d3) to the three sides from point X will always sum to 3r, regardless of location of point X within the triangle.

Table 5-1: Triangle ABC – Proof of Equality

In the figures below, each of the types of triangles based on the length of sides is shown.

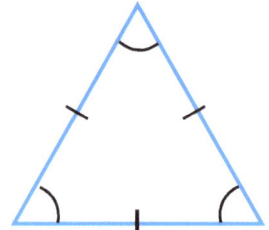 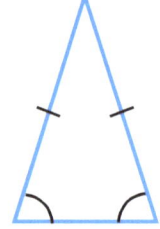 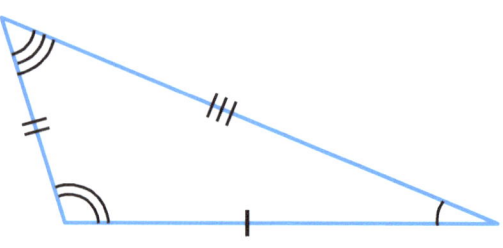

Figure 5-5: Equilateral triangle   Figure 5-6: Isosceles triangle   Figure 5-7: Scalene triangle

Based on the measurement of the angles, a triangle can be equiangular, right, oblique, acute, or obtuse. An equiangular triangle has three angles of equal measurement, equal to 60 degrees, and is a regular polygon. Every equiangular triangle is also an equilateral triangle, with all three sides having the same length. The geometric center of a regular triangle is the center of the inscribed and circumscribed circles. In the table below, formulas for a regular triangle (equilateral or equiangular) are shown.

Regular Triangle Formulas
Perimeter: $P = a + b + c$ or $3a$
Angles: $A = B = C = 60^0$; $A + B + C = 180^0$
Area: $K = 1/4 \ a^2 \ \sqrt{3}$ or $1/2 \ ah$ or $0.433012702 \times a^2$
Inradius: $r = 1/6 \ a \ \sqrt{3}$
Circumradius: $R = 1/3 \ a \ \sqrt{3}$
Altitude or Height: $h = 1/2 \ a \ \sqrt{3}$
Area of Incircle: $Kr = \pi \ r^2$ or $1/12 \ \pi \ a^2$
Area of Circumcircle: $KR = \pi \ R^2$ or $1/3 \ \pi \ a^2$

Table 5-2: Regular triangle formulas

A right triangle has one angle equal to 90 degrees or a right angle. The remaining two legs are complementary or sum to 90 degrees. The side opposite the right angle is called the hypotenuse and the other sides are called the legs. The hypotenuse is always the longest side. The distances between the midpoint of the hypotenuse and all three vertices are the same. The lengths of the sides, in relation to each other, can be determined using the Pythagorean Theorem: $a^2 + b^2 = c^2$. The sum of the squares of the two legs is equal to the square of the hypotenuse. For example, if the sides are equal to 3, 4, and 5 units, then $3^2 + 4^2 + = 5^2$ or $9 + 16 = 25$. When the lengths of two sides of a triangle are known, the length of the third side can be determined using the formula. In the figures below, four right triangles are shown.

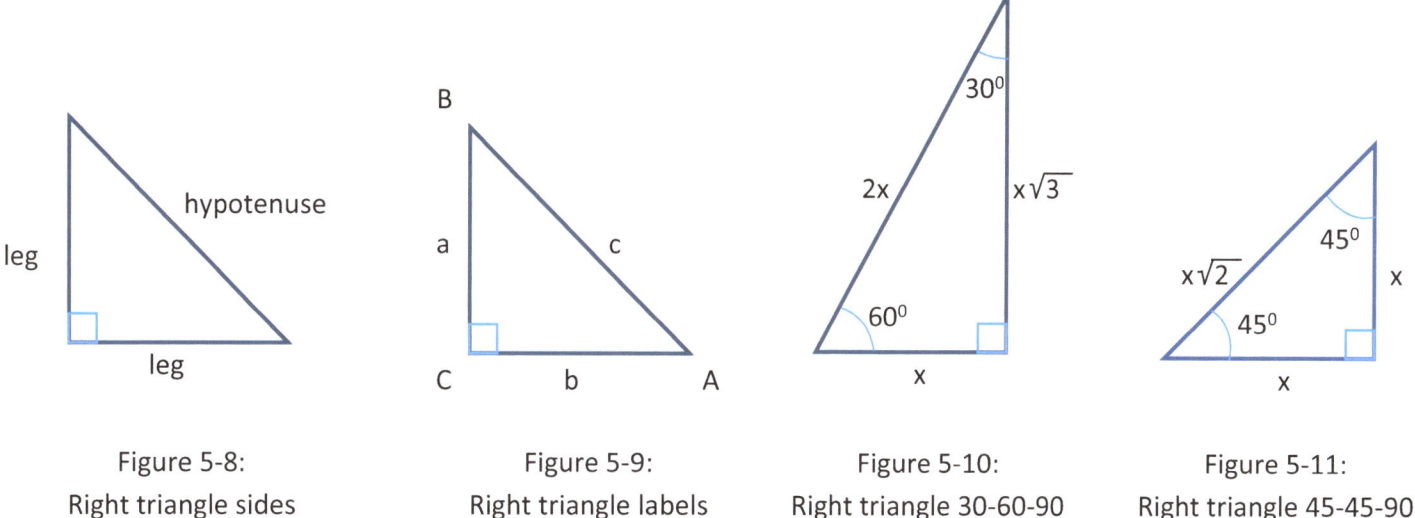

Figure 5-8:
Right triangle sides

Figure 5-9:
Right triangle labels

Figure 5-10:
Right triangle 30-60-90

Figure 5-11:
Right triangle 45-45-90

Two special or common right triangles are 30-60-90 and 45-45-90. A 30-60-90 triangle has angles of 30^0, 60^0, and 90^0. A 45-45-90 triangle has angles of 45^0, 45^0, and 90^0. Each of the two special right triangles has a constant ratio of the side lengths that can be used in calculations. For example, the length of the hypotenuse of a 30-60-90 triangle is 6 units. The legs are 3 units and $3\sqrt{3}$ units or approximately 5.19 units. In another example, the leg of a 45-45-90 triangle is 5 units. The hypotenuse is $5\sqrt{2}$ or approximately 7.07 units.

The areas of triangles in a previous example, shown again below, can be calculated.

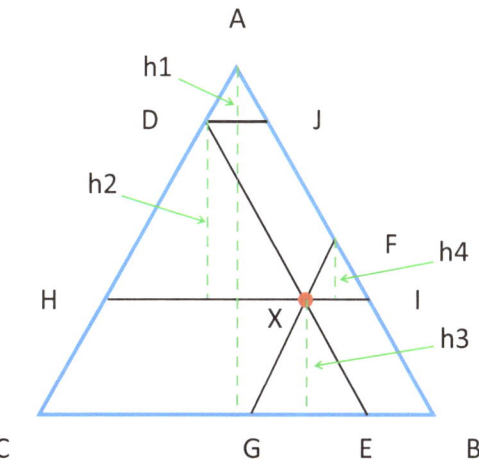

If h1 = 6, then the area of triangle ABC = 5.196152423 units.

If h2 = 3, then the area of triangle DHX = 2.598076211 units.

If h3 = 2, then the area of triangle EGX = 1.732050808 units.

If h4 = 1, then the area of triangle FIX = 0.866025404 units.

h1 = h2 + h3 + h4

Area of ABC = area of DHX + area of EGX + area of FIX

Figure 5-12: Triangle ABC – Triangle areas

Table 5-3: Triangle ABC – Area Calculations

An acute triangle has all angles measuring less than 90 degrees. An obtuse triangle has one angle measuring greater than 90 degrees. Equiangular triangles are also acute triangles. An oblique triangle does not have an angle measuring 90 degrees. All acute and obtuse triangles are also oblique triangles.

In the figures below, the types of triangles based on the measurement of the angles are shown.

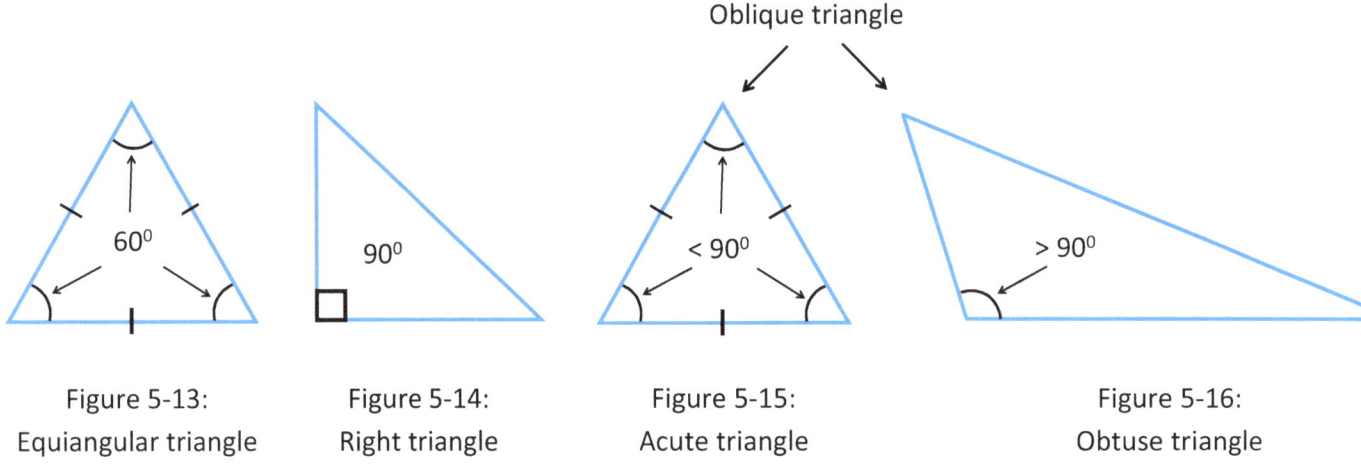

Figure 5-13:
Equiangular triangle

Figure 5-14:
Right triangle

Figure 5-15:
Acute triangle

Figure 5-16:
Obtuse triangle

• 5-3 Similar and Congruent Triangles

Numbers that have the same value are equal (=). Numbers that are almost equal are approximately equal (≈). Geometric objects, such as angles and polygons, which have the same shape and size, are congruent (≅). Objects which have the same shape but not the same size are similar (~). Equivalence is used to compare numbers, and congruence is used to compare objects.

Two triangles are similar if the corresponding angles are congruent or the corresponding sides have lengths that are in the same proportion. Similar triangles have the exact same shape, but may not be the same size. There are several theorems, or proven mathematical statements, concerning similar triangles.

Similar triangle theorem 1: If two corresponding angles of two triangles are congruent, the triangles are similar.

Similar triangle theorem 2: If two corresponding sides of two triangles are in proportion, and their included angles are congruent, the triangles are similar.

Similar triangle theorem 3: If three corresponding sides of two triangles are in proportion, the triangles are similar.

In the figures below, similar triangles are shown.

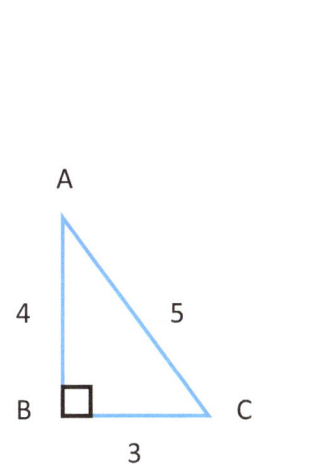 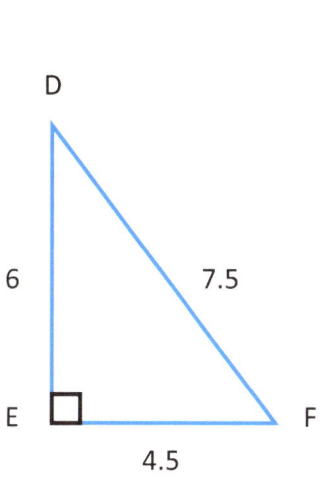 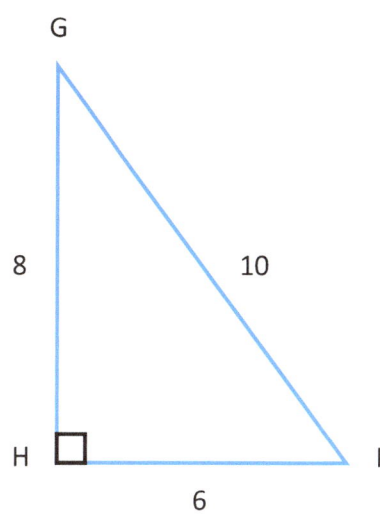

Figure 5-17: Triangle ABC   Figure 5-18: Triangle DEF   Figure 5-19: Triangle GHI

In the above figures, the triangles are similar (ABC ~ DEF ~ GHI). The ratios of the side lengths, comparing the legs of the triangles, are equal (BC/AB = EF/DE = HI/GH). The corresponding angles are congruent (A ≅ D ≅ G, B ≅ E ≅ H, C ≅ F ≅ I). If a line segment is drawn parallel to any side, connecting two sides, the resulting triangle is similar to the original triangle. In the figures below, similar triangles are created by drawing parallel lines.

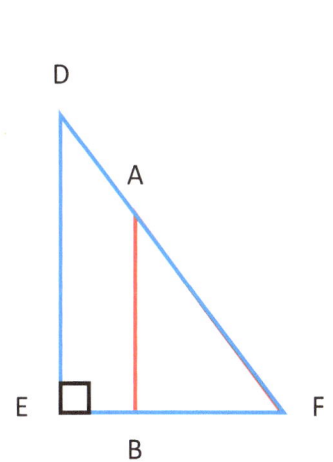 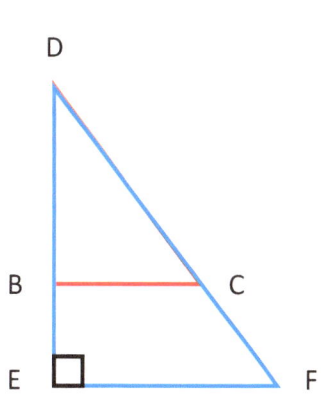 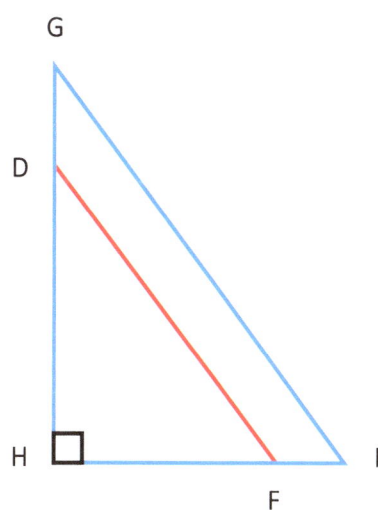

Figure 5-20:   Figure 5-21:   Figure 5-22:
Draw line segment AB   Draw line segment BC   Draw line segment DF

In the first figure above, line segment AB is drawn parallel to side DE. Triangle ABF is similar to triangle DEF. In the second figure above, line segment BC is drawn parallel to side EF. Triangle BCD is similar to triangle EFD. In the third figure above, line segment DF is drawn parallel to side GI. Triangle DFH is similar to triangle GHI.

Two triangles are congruent if all pairs of corresponding angles are congruent and all pairs of corresponding sides have the same length. This is a total of six equalities. Congruent triangles have the exact same shape and size. If triangles are congruent, then they are also similar. The following equality conditions are sufficient to prove congruency for a pair of triangles.

1. (SAS): Two sides in a triangle have the same length as two sides in the other triangle, and the included angles have the same measure.
2. (ASA): Two angles in a triangle have the same measure as two angles in the other triangle, and the included side has the same length.
3. (SSS): Each side of a triangle has the same length as a corresponding side of the other triangle.
4. (AAS): Two angles in a triangle have the same measure as two angles in the other triangle, and a corresponding non-included side has the same length.
5. (SSA): Two sides in a triangle have the same length as two sides in the other triangle, and the corresponding non-included angles have the same measure, but only if the angle is opposite the longer of the two sides.
6. (HL): The hypotenuse and a leg in a right triangle have the same length as those in the other right triangle. This is a variation of SSA.
7. (HA): The hypotenuse and an acute angle in a right triangle have the same length and measure, respectively, as those in the other right triangle. This is a variation of AAS.
8. (LL): The legs of a right triangle have the same length as the legs of the other right triangle. This is a variation of SAS.
9. (LA): A leg and an acute angle in a right triangle have the same length and measure, respectively, as those in the other right triangle. This is a variation of ASA and AAS.

When two triangles are congruent, all six pairs of corresponding parts are congruent. The parts of two triangles that have the same measurements are referred to as corresponding parts. This means that corresponding parts of congruent triangles are congruent (CPCTC). In the figures below, congruent triangles are shown.

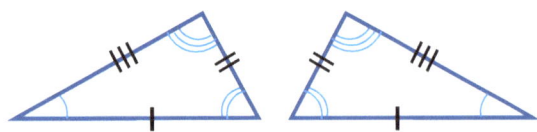  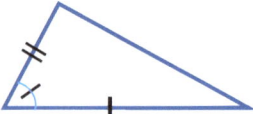

Figure 5-23: Congruent triangles - CPCTC   Figure 5-24: Congruent triangles - SAS

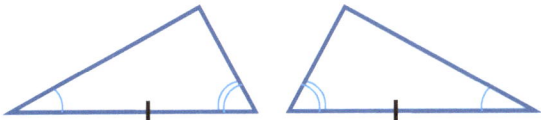

Figure 5-25: Congruent triangles - ASA

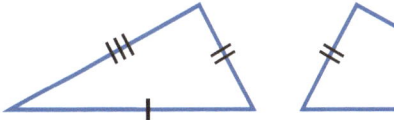

Figure 5-26: Congruent triangles - SSS

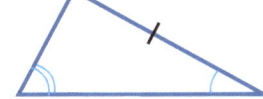

Figure 5-27: Congruent triangles - AAS

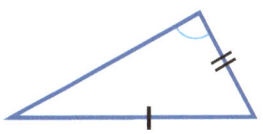

Figure 5-28: Congruent triangles - SSA

Figure 5-29: Congruent triangles - HL

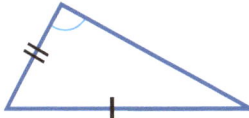

Figure 5-30: Congruent triangles - HA

Figure 5-31: Congruent triangles - LL

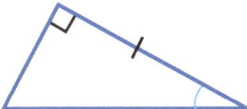

Figure 5-32: Congruent triangles – LA (ASA)

Figure 5-33: Congruent triangles – LA (AAS)

5-4 Points, Lines, and Circles of Triangles

There are many special points associated triangles which satisfy unique properties. Some of the most common points are explained in this section.

The **circumcircle** is a circle outside of a polygon that intersects each of the vertices. The center of this circle is the **circumcenter**. The **perpendicular bisector** of a side of a triangle is a straight line passing through the midpoint of the side and being perpendicular to it, forming a right angle with it. The three perpendicular bisectors meet in a single point, the circumcenter. The location of the circumcenter of a triangle depends on the type of triangle. For right triangle it is located on one of the sides. For an acute triangle it is located inside the triangle. For an obtuse triangle it is located outside of the triangle. The circumcenter is equidistant from the three vertices and the common distance is the radius of the circumcircle.

An altitude of a triangle is a straight line through a vertex and perpendicular to the opposite side. The three altitudes intersect in a single point, called the orthocenter of the triangle. The orthocenter lies inside the triangle only if the triangle is acute.

The incircle is a circle inside of a polygon that touches each of the sides. The center of this circle is the incenter. An angle bisector of a triangle is a straight line through a vertex which cuts the corresponding angle in half. The three angle bisectors intersect in a single point, the incenter. The incenter is equidistant from the three sides and the common distance is the radius of the incircle.

In the figures below, the circumcenter, orthocenter, and incenter of triangles are shown.

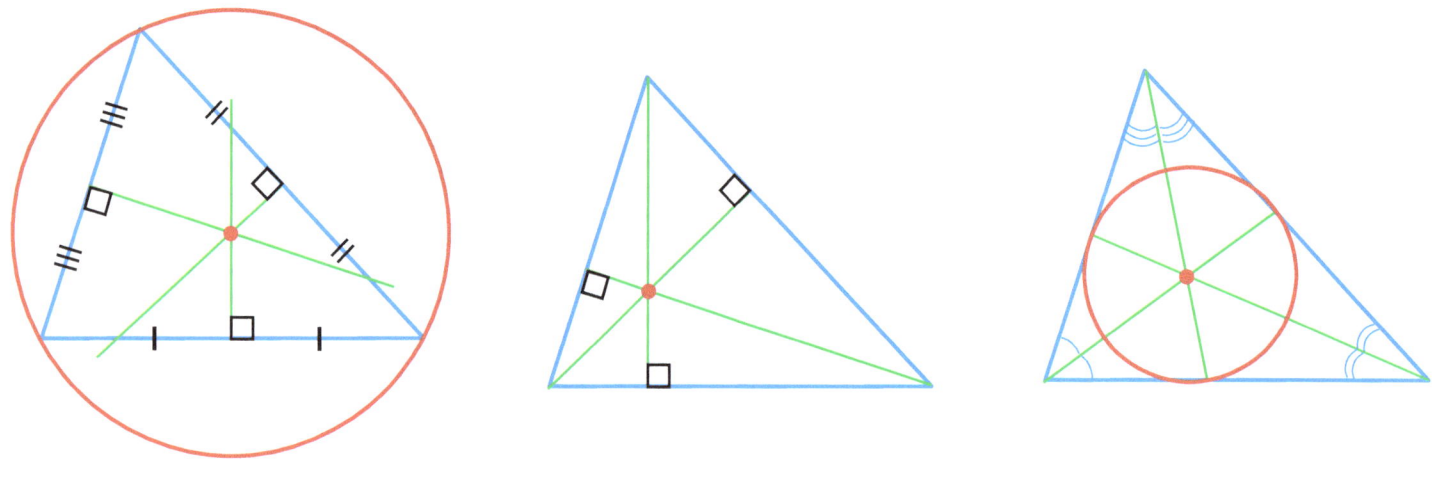

Figure 5-34:
Circumcenter of a triangle

Figure 5-35:
Orthocenter of a triangle

Figure 5-36:
Incenter of a triangle

A median of a triangle is a straight line through a vertex and the midpoint of the opposite side, and divides the triangle into two equal areas. The three medians intersect in a single point, the centroid. The centroid cuts every median in a 2:1 ratio. The distance between a vertex and the centroid is twice the distance between the centroid and the midpoint of the opposite side. The centroid is also the center of mass or gravity of an object. If a triangle is cut out of a piece of cardboard, it would balance at the centroid. One of the many interesting properties of the centroid of a triangle is that it is the unique point D for which the three triangles BCD, CAD, ABD all have the same area. This is shown in the first figure below.

The Fermat point is a point of a triangle that is a minimum distance from the three vertices. There are two methods for finding the Fermat point. Each starts by constructing an equilateral triangle on each side of the triangle. For the first method, draw line segments connecting the vertices of the triangle and the opposite equilateral triangles. The intersection of the lines is the Fermat point. For the second method, draw circumcircles around each of the equilateral triangles. The intersection of the circumcircles is the Fermat point. In an acute triangle, the angle formed by the segments connecting any two vertices with the Fermat point is 120 degrees.

In the figures below, the centroid and Fermat point of a triangle are shown.

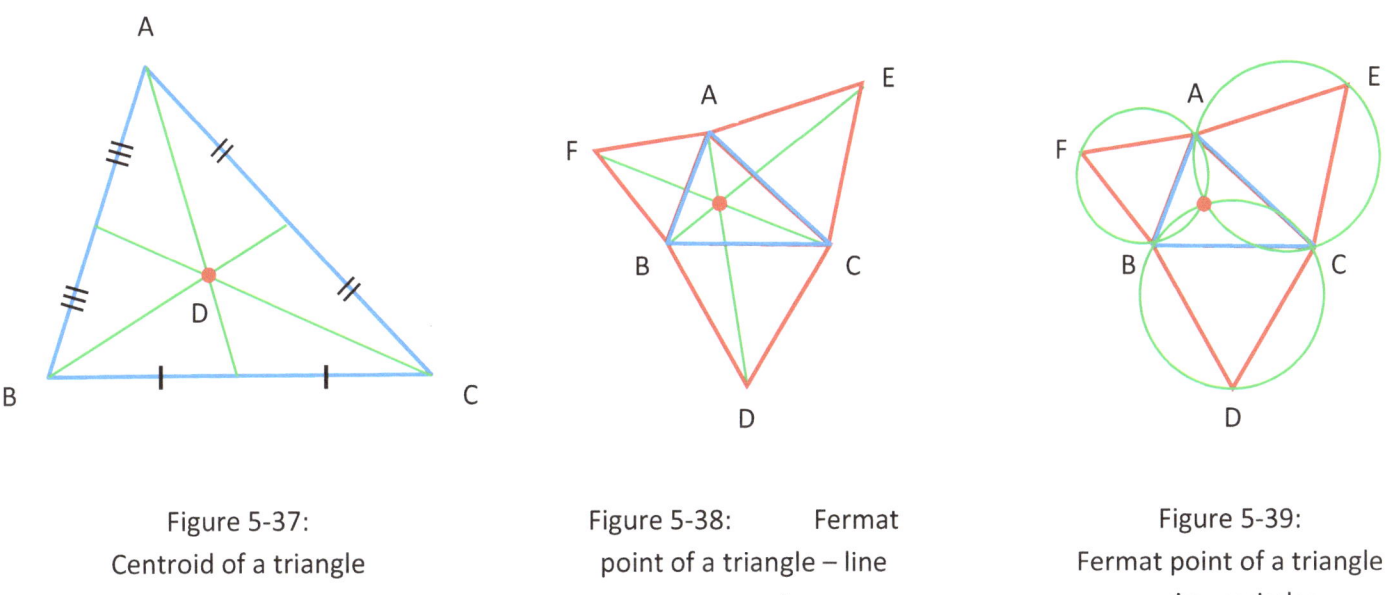

Figure 5-37:
Centroid of a triangle

Figure 5-38:   Fermat point of a triangle – line segments

Figure 5-39:
Fermat point of a triangle – circumcircles

The Morley triangle is an equilateral triangle formed by the three points of intersection of the adjacent angle trisectors. The Morley center is the intersection point of the line segments connecting the Morley triangle vertices with the opposite original triangle vertices. The intersection of each line segment with the side of the Morley triangle will always form a right angle. It should be noted that while the Morley center may visually appear to coincide with the center of the inscribed circle of the original triangle, they are not exactly coincident. The Morley circle is the circumcircle of the Morley triangle.

In the figures below, Morley triangles of triangles are shown.

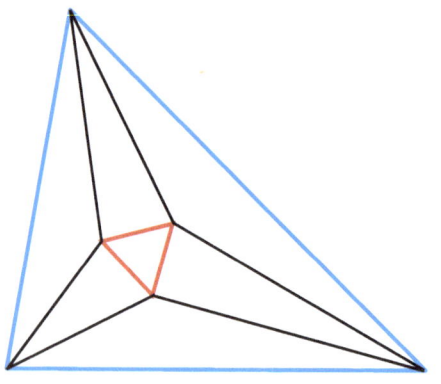
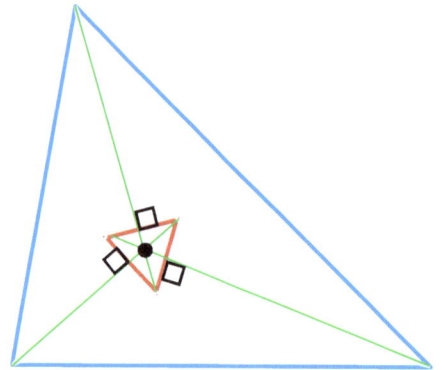
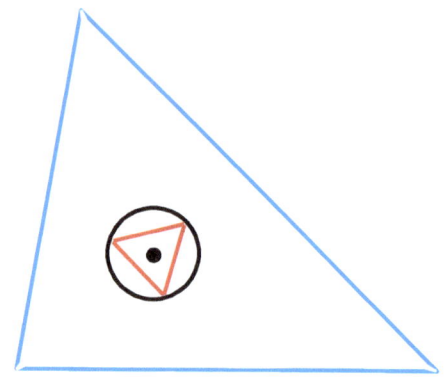

Figure 5-40: Morley triangle   Figure 5-41: Morley center   Figure 5-42: Morley circle

5-5 Triangle Rigidity

The triangle is the only rigid polygon. This **rigidity** is defined by two statements. First, if the lengths of the sides are known, then the angles can not be changed by shifting the sides. This means the only way to change the angles is by changing the length of the sides. Second, the length of less than all of the sides can not be changed without changing the angles. This means if the length of only one or two sides is changed, the angles must change. Also, the lengths of all three sides must be changed by a common factor, if the angles are not to be changed. In the figures below, the first rigidity statement is shown, that sides of a triangle can not be shifted.

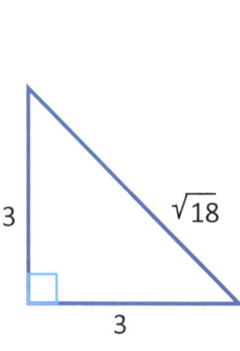
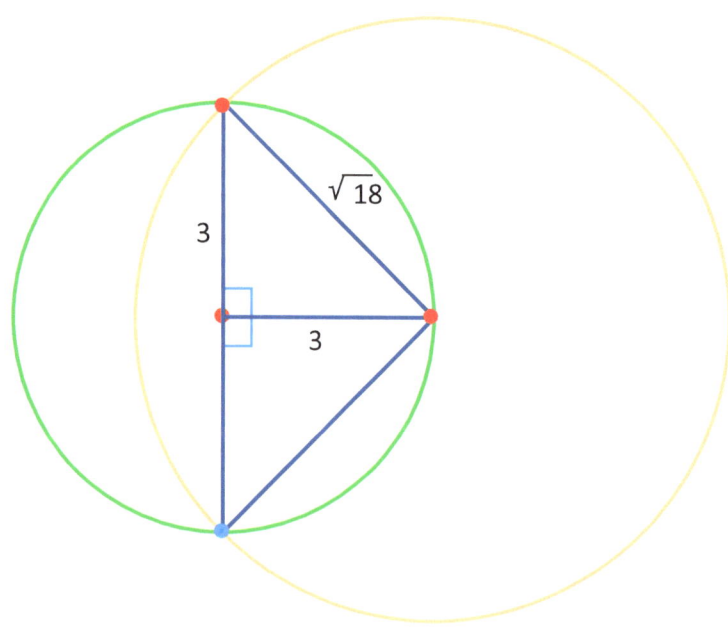

Figure 5-43: Sample triangle – 1   Figure 5-44: Triangle – shifting sides

The first figure above is a sample triangle. The second figure above is a triangle with shifting sides. If the horizontal side is fixed in position, then the vertical side can be rotated along the smaller circle and length of that side will remain unchanged. The diagonal side or hypotenuse can be rotated along the larger circle and the length of that side will remain unchanged. The intersection of the two circles indicates the only locations where each of the three side lengths remains unchanged. The triangle can be flipped vertically, but the sides can not be rotated or shifted.

In the figures below, the first rigidity statement does not apply to quadrilaterals, as the sides can be shifted.

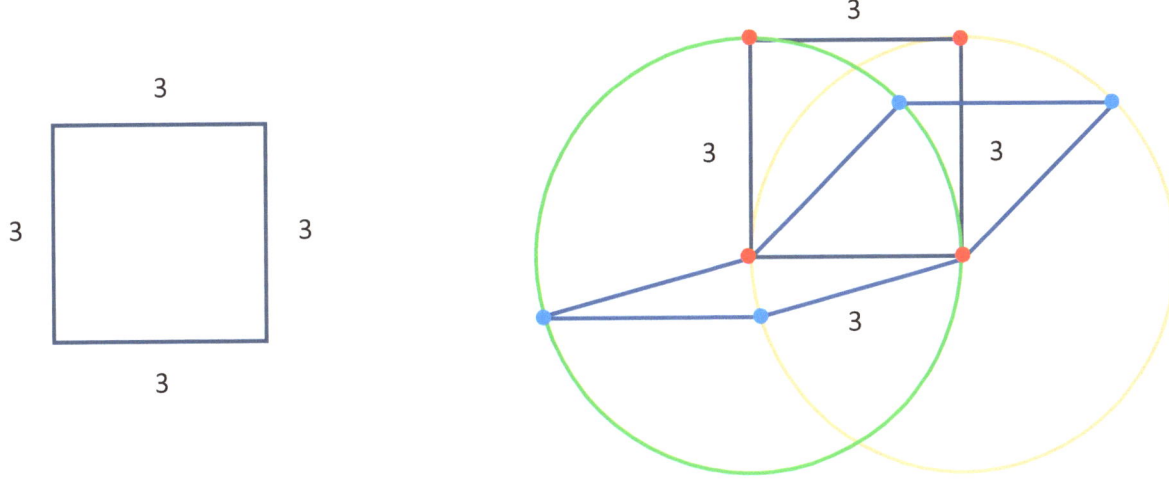

Figure 5-45: Sample quadrilateral – 1   Figure 5-46: Quadrilateral – shifting sides

The first figure above is a sample quadrilateral. The second figure above is a quadrilateral with shifting sides. If the bottom side is fixed in position, then the left and right sides can be rotated along the two circles and the top side will remain parallel to the bottom side. The length of each of the four sides will remain unchanged. The sides of the quadrilateral can be shifted to any angles.

In the figures below, the second rigidity statement is shown, that the length of less than all of the sides can not be changed without changing the angles.

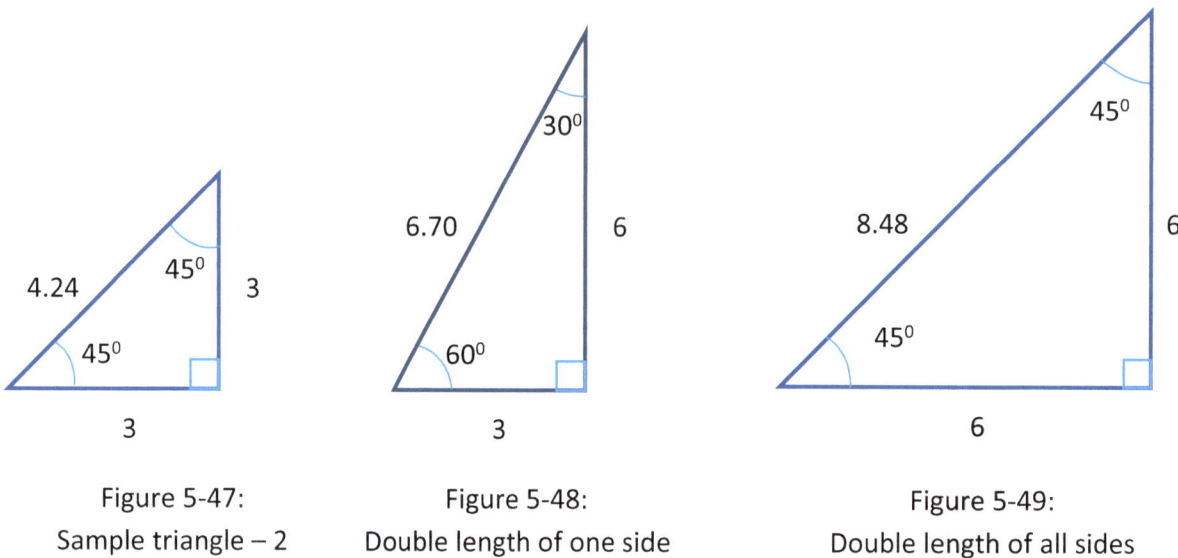

Figure 5-47:
Sample triangle – 2

Figure 5-48:
Double length of one side

Figure 5-49:
Double length of all sides

The first figure above, the sample triangle angles are 45, 45, and 90 degrees. In the second figure above, one side length is doubled and the angles of the triangle change to 30, 60, and 90 degrees. In the third figure above, all side lengths are doubled and the angles remain as 45, 45, and 90 degrees.

In the figures below, the second rigidity statement does not apply to quadrilaterals, as that the length of less than all of the sides can be changed without changing the angles.

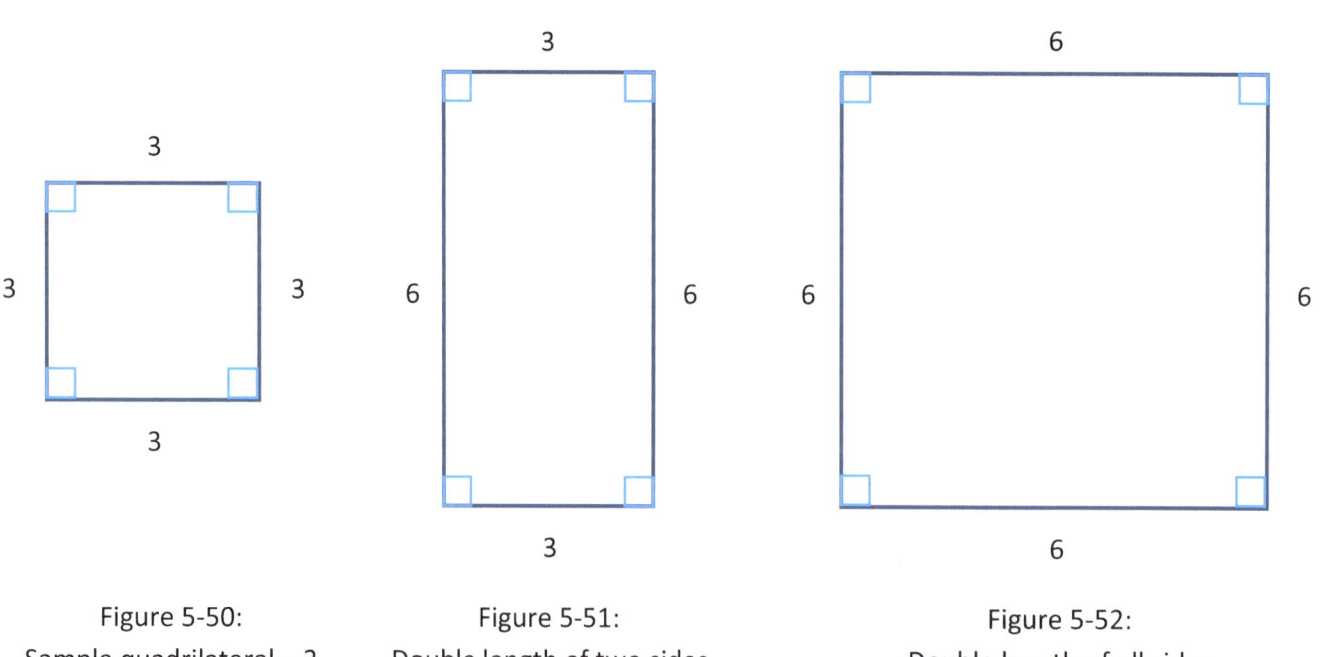

Figure 5-50:
Sample quadrilateral – 2

Figure 5-51:
Double length of two sides

Figure 5-52:
Double length of all sides

The first figure above is a sample quadrilateral. The second figure above is a quadrilateral with the length of two opposite sides doubled. The third figure above is a quadrilateral with the length of all four sides doubled. In each of the quadrilaterals, all four angles remain as 90 degrees.

5-6 Summary

A triangle is a polygon with three sides and three angles. There are many types of triangles based on the lengths of the sides and the measurements of the angles. A triangle with vertices A, B, and C is denoted $\triangle ABC$. All triangles are convex and bicentric. Bicentric means a polygon possesses both an incircle and a circumcircle. The sum of all angles in any triangle is 180 degrees. Triangles can be divided into types based on the length of sides or the measurement of the angles. A triangle can be a combination of types. Based on the length of the sides, a triangle can be equilateral, isosceles, or scalene. An equilateral triangle has three sides of equal length and is a regular polygon. An isosceles triangle has at least two side of equal length. A scalene triangle has all three sides of unequal length.

Based on the measurement of the angles, a triangle can be equiangular, right, oblique, acute, or obtuse. An equiangular triangle has three angles of equal measurement, equal to 60 degrees, and is a regular polygon. A right triangle has one angle equal to 90 degrees or a right angle. An acute triangle has all angles measuring less than 90 degrees. An obtuse triangle has one angle measuring greater than 90 degrees. Equiangular triangles are also acute triangles. An obtuse triangle does not have an angle measuring 90 degrees. All acute and obtuse triangles are also oblique triangles.

Two triangles are similar if the corresponding angles are congruent or the corresponding sides have lengths that are in the same proportion. Similar triangles have the exact same shape, but may not be the same size. Two triangles are congruent if all pairs of corresponding angles are congruent and all pairs of corresponding sides have the same length. This is a total of six equalities. Congruent triangles have the exact same shape and size. If triangles are congruent, then they are also similar. When two triangles are congruent, all six pairs of corresponding parts are congruent. The parts of two triangles that have the same measurements are referred to as corresponding parts. This means that corresponding parts of congruent triangles are congruent (CPCTC).

There are many special points associated triangles which satisfy unique property. The circumcircle is a circle outside of a polygon that intersects each of the vertices. The center of this circle is the circumcenter. The perpendicular bisector of a side of a triangle is a straight line passing through the midpoint of the side and being perpendicular to it, forming a right angle with it. The three perpendicular bisectors meet in a single point, the circumcenter.

An altitude of a triangle is a straight line through a vertex and perpendicular to the opposite side. The three altitudes intersect in a single point, called the orthocenter of the triangle. The incircle is a circle inside of a polygon that touches each of the sides. The center of this circle is the incenter. An angle bisector of a triangle is a straight line through a vertex which cuts the corresponding angle in half. The three angle bisectors intersect in a single point, the incenter. The incenter is equidistant from the three sides and the common distance is the radius of the incircle.

A median of a triangle is a straight line through a vertex and the midpoint of the opposite side, and divides the triangle into two equal areas. The three medians intersect in a single point, the centroid. The Fermat point is a point of a triangle that is a minimum distance from the three vertices. The Morley triangle is an equilateral triangle formed by the three points of intersection of the adjacent angle trisectors.

The triangle is the only rigid polygon. This rigidity is defined by two statements. First, if the lengths of the sides are known, then the angles can not be changed by shifting the sides. This means the only way to change the angles is by changing the length of the sides. Second, the length of less than all of the sides can not be changed without changing the angles. This means if the length of only one or two sides is changed, the angles must change. Also, the lengths of all three sides must be changed by a common factor, if the angles are not to be changed.

- ## Section 2 – Quadrilaterals

- ### 5-1 Introduction

A ==quadrilateral== is a polygon with four sides and four angles. The sum of the interior angles is 360 degrees. There are three types of quadrilaterals: convex, concave, and complex. In the figures below, the three types of quadrilaterals are shown.

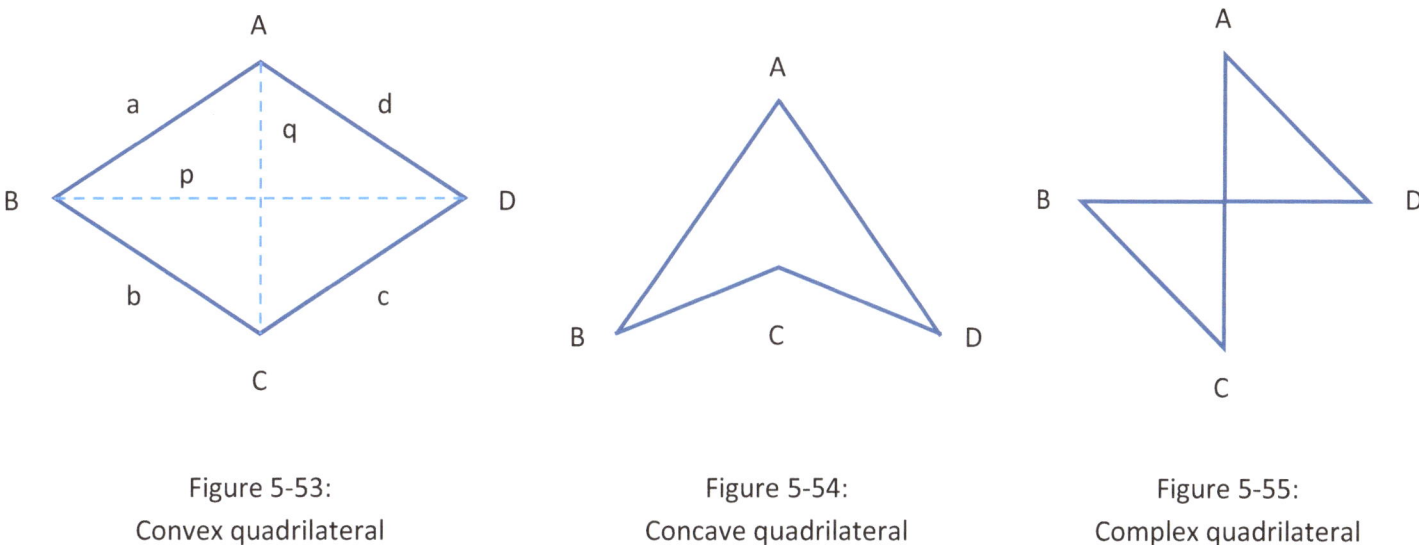

Figure 5-53:
Convex quadrilateral

Figure 5-54:
Concave quadrilateral

Figure 5-55:
Complex quadrilateral

In the first figure above, a ==convex quadrilateral== is shown. The four sides are labeled as a, b, c, and d. The four angles are labeled as A, B, C, and D. The diagonals, lines connecting the opposite vertices, are labeled as p and q. In the second figure above, the ==concave quadrilateral== is called a dart or arrow. It has two pairs of adjacent equal sides. In the third figure above, the lines of the sides cross resulting in a ==complex quadrilateral==. A complex quadrilateral is self-intersecting because a pair of non-adjacent sides intersects. This shape is called a butterfly, bow tie, or hour glass. It has two 45-45-90 degree triangles connected at a common vertex.

- ### 5-8 Convex Quadrilaterals

Quadrilaterals are prevalent shapes and can be classified based on the special relationships of the side lengths and angle measurements. There are seven types of convex quadrilaterals based on special properties. In the table below, quadrilateral names and properties are shown.

Quadrilateral Name	Properties
Kite	2 pairs of adjacent equal sides
Trapezoid	1 pair of parallel sides
Isosceles Trapezoid	1 pair of parallel sides and 1 pair of equal sides
Parallelogram	2 pairs of parallel sides
Rhombus	2 pairs of parallel sides and 4 equal sides
Rectangle	2 pairs of parallel sides and all angles equal to 90 degree angles
Square	2 pairs of parallel sides, all angles equal to 90 degrees, and 4 equal sides

Table 5-4: Quadrilateral types

In the figures below, the seven types of convex quadrilaterals are shown.

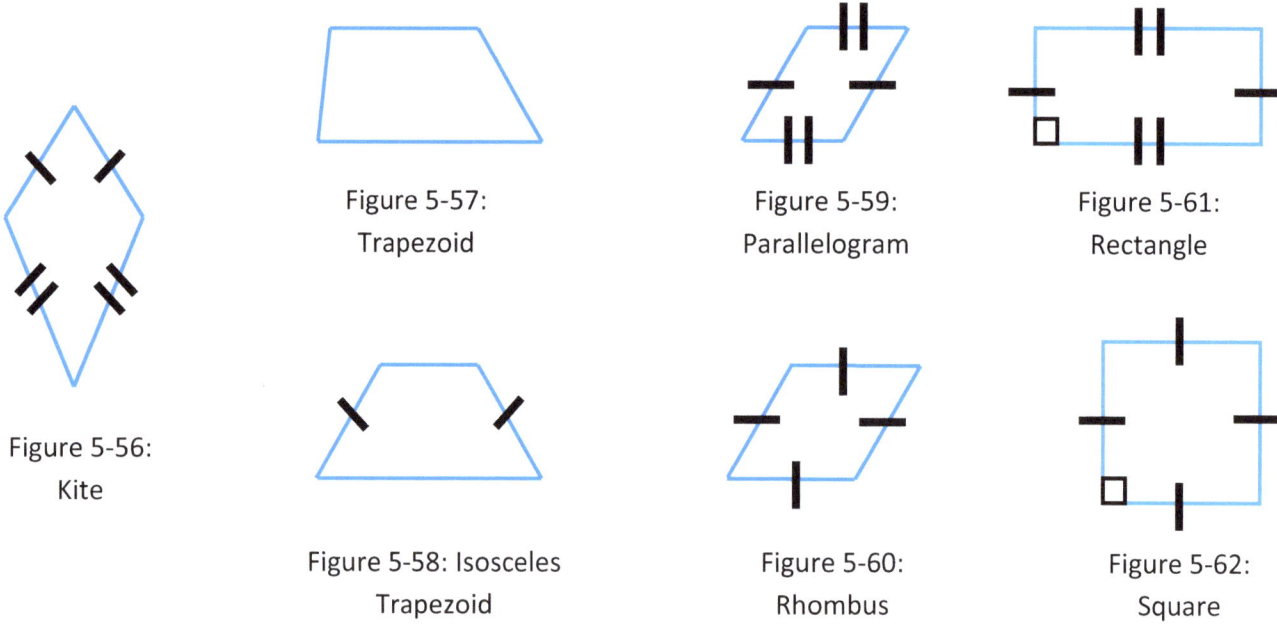

Figure 5-56: Kite

Figure 5-57: Trapezoid

Figure 5-58: Isosceles Trapezoid

Figure 5-59: Parallelogram

Figure 5-60: Rhombus

Figure 5-61: Rectangle

Figure 5-62: Square

In the figures above, the seven types of quadrilaterals are shown. The quadrilaterals share properties, such as parallel sides or equal sides. Four of the shapes have 2 pairs of parallel sides, including the square. Two of the shapes have all angles equal to 90 degrees, including the square. Two of the shapes have 4 equal sides, including the square. A combination of the three previous statements uniquely identifies a square. The square is the only regular quadrilateral. Some quadrilaterals may not have any special properties, and generally do not have parallel sides, equal sides, or congruent angles. These quadrilaterals are placed in an eighth category called scalene quadrilaterals.

Quadrilaterals can be divided according to their properties. Orthodiagonal means the diagonals form perpendicular or right angles. Cyclic means the quadrilateral can be circumscribed. Inscriptible means the quadrilateral can be inscribed. A quadrilateral that is both cyclic and tangential (inscribed so the circle is tangent or touching each side) is called a bicentric quadrilateral. A kite is cyclic only if it has two right angles. In the table below, additional properties and quadrilateral names are shown.

Property	Quadrilaterals
Orthodiagonal	Kite, Rhombus, Square
Cyclic	Isosceles Trapezoid, Kite, Rectangle, Square
Inscriptible	Kite, Rhombus, Square
Bicentric	Square
One pair of parallel sides	Isosceles Trapezoid, Trapezoid
Two pairs of parallel sides	Parallelogram, Rectangle, Rhombus, Square
One pair of equal sides	Isosceles Trapezoid
Two pairs of equal sides	Kite, Parallelogram, Rectangle
Equilateral	Rhombus, Square
Equiangular	Rectangle, Square
Equal opposite angles	Kite, Parallelogram, Rhombus

Table 5-5: Quadrilateral properties

• 5-9 Quadrilateral Formulas

An equation can be used to determine if the diagonals of a convex quadrilateral are perpendicular. The four sides are labeled as a, b, c, and d. The equation is: $a^2 + c^2 = b^2 + d^2$. For example, if a = 2, b = 3, c = 3, and d = 2, then $4 + 9 = 9 + 4$. The diagonals are perpendicular.

The area of a cyclic quadrilateral can be calculated using Brahmagupta's formula. The semiperimeter is $s = (a + b + c + d) / 2$. The area formula is $k = \sqrt{((s - a)(s - b)(s - c)(s - d))}$. The area formula can also be written as $k = 1/4 \sqrt{((a+b+c-d)(a+b-c+d)(a-b+c+d)(-a+b+c+d))}$.

If a cyclic quadrilateral is also orthodiagonal, the perpendicular from any side through the point or intersection of the diagonals bisects the other side. Also, the distance from the circumcenter to any side equals half the length of the opposite side.

The tables below show formulas for calculating quadrilateral measurements.

General Quadrilateral Formulas

$s = (a + b + c + d) / 2$
$K = 1/4 \sqrt{4p^2q^2 - (b^2 + d^2 - a^2 - c^2)^2}$

Cyclic Quadrilateral Formulas

$K = \sqrt{(s-a)(s-b)(s-c)(s-d)}$
$p = \sqrt{((ac+bd)(ab+cd))/(ad+bc)}$
$q = \sqrt{((ac+bd)(ad+bc))/(ab+cd)}$
$R = 1/4 \sqrt{((ac+bd)(ad+bc)(ab+cd))/((s-a)(s-b)(s-c)(s-d))}$

Cyclic Inscriptable Quadrilateral Formulas

$A + C = B + D = 180^0$
$K = \sqrt{abcd}$
$r = (\sqrt{abcd})/s$
$R = 1/4 \sqrt{((ac+bd)(ad+bc)(ab+cd))/(abcd)}$

Kite Formulas

$K = (pq)/2$
$B = D$

Parallelogram Formulas

$A = C, B = D, A + B = 180^0$
$K = bh$

Rectangle Formulas

$A = B = C = D = 90^0$
$K = ab$
$p = \sqrt{a^2 + b^2}$

Rhombus Formulas

$p^2 + q^2 = 4a^2$
$K = 1/2\ pq$

Trapezoid Formulas

$m = 1/2\ (a + b)$
$K = 1/2\ (a + b)\ h = mh$

Table 5-6: Quadrilateral formulas

The figures below are used with the table above.

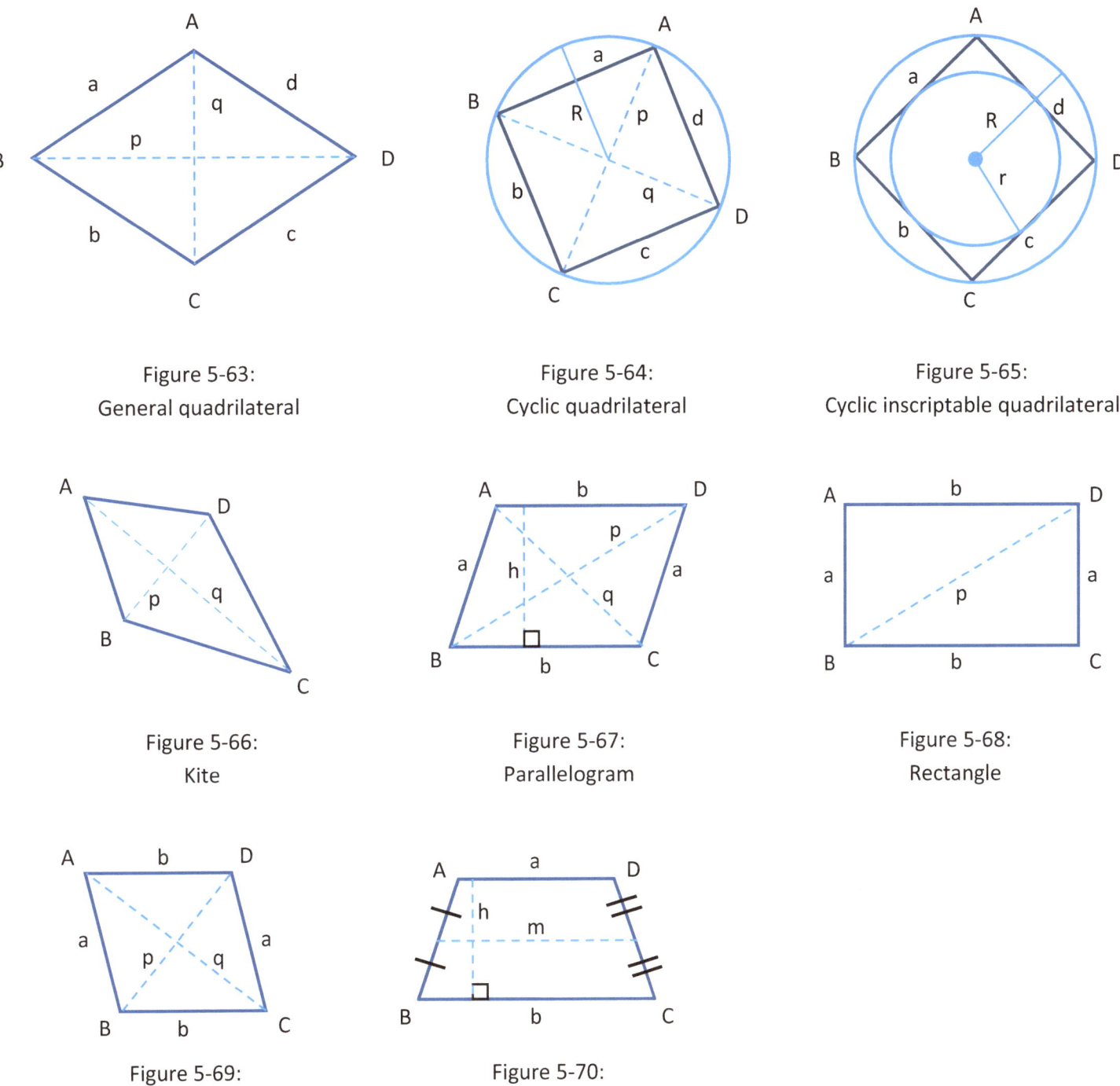

Figure 5-63: General quadrilateral

Figure 5-64: Cyclic quadrilateral

Figure 5-65: Cyclic inscriptable quadrilateral

Figure 5-66: Kite

Figure 5-67: Parallelogram

Figure 5-68: Rectangle

Figure 5-69: Rhombus

Figure 5-70: Trapezoid

5-10 Summary

A Quadrilateral is a polygon with four sides and four angles. The sum of the interior angles is 360 degrees. There are three types of quadrilaterals: convex, concave, and complex. Quadrilaterals are prevalent shapes and can be classified based on the special relationships of the side lengths and angle measurements. There are seven types of convex quadrilaterals based on special properties. These quadrilaterals are the kite, trapezoid, isosceles trapezoid, parallelogram, rhombus, rectangle, and square. The quadrilaterals share properties, such as parallel sides or equal sides. Some quadrilaterals may not have any special properties, and generally do not have parallel sides, equal sides, or congruent angles. These quadrilaterals are placed in an eighth category called scalene quadrilaterals.

Quadrilaterals can be divided according to their properties. Orthodiagonal means the diagonals form perpendicular or right angles. Cyclic means the quadrilateral can be circumscribed. Inscriptible means the quadrilateral can be inscribed. A quadrilateral that is both cyclic and tangential (inscribed so the circle is tangent or touching each side) is called a bicentric quadrilateral. A kite is cyclic only if it has two right angles.

An equation can be used to determine if the diagonals of a convex quadrilateral are perpendicular. The area of a cyclic quadrilateral can be calculated using Brahmagupta's formula. If a cyclic quadrilateral is also orthodiagonal, the perpendicular from any side through the point or intersection of the diagonals bisects the other side. Also, the distance from the circumcenter to any side equals half the length of the opposite side.

CHAPTER 5

Chapter Test

Grading Scale: One point for each correct answer.

Excellent = 74-82, Good = 66-73, Average = 58-65, Fair = 50-57, Poor = 0-49

Section 1 – Triangles

5-2 Triangle Types

Calculate values for a regular triangle with side = 5. (Round to the nearest thousandth.)

1. Area (K) _____ square units
2. Inradius (r) _____ units
3. Circumradius (R) _____ units
4. Altitude or Height (h) _____ units
5. Area of Incircle (Kr) _____ square units
6. Area of Circumcircle (KR) _____ square units

Calculate right triangle sides using the Pythagorean Theorem. (Round to the nearest thousandth.)

7. C, with A = 2 and B = 4 _____ units
8. B, with A = 5 and C = 10 _____ units
9. A, with B = 4 and C = 8 _____ units
10. C, with A = 2 and B = 2 _____ units
11. B, with A = 6 and C = 10 _____ units
12. A, with B = 10 and C = 15 _____ units

Calculate sides A and C in a 30-60-90 right triangle. (Round to the nearest thousandth.)

13. B = 2 _____ units _____ units
14. B = 4 _____ units _____ units
15. B = 6 _____ units _____ units
16. B = 8 _____ units _____ units

Calculate sides A and C in a 45-45-90 right triangle. (Round to the nearest thousandth.)

17. B = 3 _____ units _____ units
18. B = 6 _____ units _____ units
19. B = 9 _____ units _____ units
20. B = 12 _____ units _____ units

5-3 Similar and Congruent Triangles

Determine if two triangles are similar or congruent, then list the theorem.

1. Two corresponding angles of two triangles are congruent.

    Similar or congruent _____   theorem _____

2. A leg and an acute angle in a right triangle have the same length and measure, respectively, as those in the other right triangle.

    Similar or congruent _____   theorem _____

3. Two angles in a triangle have the same measure as two angles in the other triangle, and a corresponding non-included side has the same length.

    Similar or congruent _____   theorem _____

4. Each side of a triangle has the same length as a corresponding side of the other triangle.

    Similar or congruent _____   theorem _____

5. Three corresponding sides of two triangles are in proportion.

    Similar or congruent _____   theorem _____

6. The hypotenuse and a leg in a right triangle have the same length as those in the other right triangle.

    Similar or congruent _____   theorem _____

7. Two sides in a triangle have the same length as two sides in the other triangle, and the included angles have the same measure.

    Similar or congruent _____   theorem _____

8. The hypotenuse and an acute angle in a right triangle have the same length and measure, respectively, as those in the other right triangle.

    Similar or congruent _____   theorem _____

9. Two sides in a triangle have the same length as two sides in the other triangle, and the corresponding non-included angles have the same measure, but only if the angle is opposite the longer of the two sides.

    Similar or congruent _____   theorem _____

10. Two corresponding sides of two triangles are in proportion, and their included angles are congruent.

   Similar or congruent _____   theorem _____

11. The hypotenuse and an acute angle in a right triangle have the same length and measure, respectively, as those in the other right triangle.

   Similar or congruent _____   theorem _____

12. Each side of a triangle has the same length as a corresponding side of the other triangle.

   Similar or congruent _____   theorem _____

5-4 Points, Lines, and Circles of Triangles

Match definitions and terms.

   A = Circumcenter   B = Orthocenter   C = Incenter

   D = Centroid   E = Fermat Point   F = Morley Triangle

1. Minimum distance from three vertices. ____
2. Intersection of three altitudes. ____
3. Intersection of three medians. ____
4. Intersection of three perpendicular bisectors. ____
5. Intersection of three adjacent angle trisectors. ____
6. Intersection of three angle bisectors. ____

5.5 – Triangle Rigidity

Mark as True or False.

1. Triangles are rigid polygons. ____
2. In a rigid polygon, the angles can not be changed by shifting the sides. ____
3. In a rigid polygon, sides can be changed without changing the angles. ____
4. Quadrilaterals are rigid polygons. ____

Section 1 – Quadrilaterals

5-8 Convex Quadrilaterals

Match definitions and terms.

A = Kite   B = Trapezoid   C = Isosceles Trapezoid   D = Parallelogram

E = Rhombus F = Rectangle   G = Square   H = Scalene Quadrilateral

1. No special properties, such a parallel sides, equal sides, or congruent angles. ____
2. Bicentric ____
3. Cyclic ____ ____ ____ ____
4. Orthodiagonal ____ ____ ____
5. Two pairs of equal sides. ____ ____ ____
6. One pair of parallel sides. ____ ____
7. All angles equal to 90 degrees. ____ ____
8. Inscriptible ____ ____ ____

5-9 Quadrilateral Formulas

Calculate the area of quadrilaterals in square units. (Round to the nearest thousandth.)

1. General quadrilateral with p = 6, q = 8, a = 5, b = 5, c = 5, d = 5 _____
2. Cyclic quadrilateral with a = 2, b = 5, c = 5, d = 8 _____
3. Cyclic inscriptable quadrilateral with a = 3, b = 3, c = 5, d = 5 _____
4. Kite quadrilateral with p = 5, q = 10 _____
5. Parallelogram quadrilateral with b = 6, h = 3 _____
6. Rectangle quadrilateral with a = 5, b = 6 _____
7. Rhombus quadrilateral with p = 8, q = 8 _____
8. Trapezoid quadrilateral with a = 2, b = 8, h = 4 _____

CHAPTER 6: Polyhedron

• 6-1 Introduction

A polyhedron is a three-dimensional shape bounded by faces (polygons), edges (line segments), and vertices (points). Since a polyhedron is comprised of polygons, it has flat faces and straight edges. The word polyhedron comes from the Greek πολύεδρον, as *poly-* (stem of πολύς, "many") + *-hedron* (form of έδρα, "seat"). The plural of polyhedron is "polyhedra" (or sometimes "polyhedrons").

To exclude abnormal cases such as T-intersections and pyramids touching at a vertex, there are two conditions that must be satisfied when constructing a polyhedron. First, every side of every polygon belongs to just one other polygon. Second, the faces that share a vertex form a chain of polygons in which every pair of consecutive polygons share a side.

A polyhedron is an enclosed three-dimensional object. There are three definitions of a polyhedron based on bounding characteristics. The first describes a polyhedron as only the edges of the polygons. There is no surface area or internal volume. It resembles a wire frame. The second states a polyhedron is a boundary between the internal and external areas formed by the faces of polygons. The volume of the internal space is not included. It is a hollow shape. The third includes the polygon surfaces and the internal volume. A polyhedron is a three-dimensional solid. When describing a polyhedron in general, the third definition is almost always used. The three polyhedron definitions are shown in the figures below.

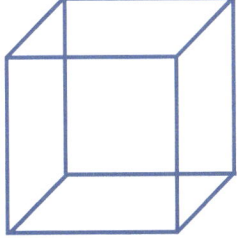

Figure 6-1: Wire Frame

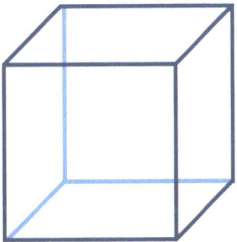

Figure 6-2: Hollow Shape

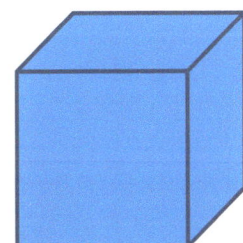

Figure 6-3: Solid Shape

6-2 Polyhedron Name Conventions

The name of a polyhedron can be based on several factors, such as the number of faces, the polygon shape of the faces, and the convex or concave surface of the faces. A polyhedron is convex if its surface does not intersect itself and a line segment drawn to connect any two points of the polyhedron is contained in the interior or on the surface. A polyhedron is concave if it has a hole or indentation and is complex if the surface is not uniformly flat or is like a star. Three types of polyhedron are shown in the figures below.

Figure 6-4:
Simple Convex
Dodecahedron

Figure 6-5:
Complex Concave
Star Dodecahedron

Figure 6-6:
Complex Convex
Star Tetrahedron

A polyhedron is classified and named according to the number and type of faces. The first polyhedra have names similar to polygons, with a Greek prefix for the number of sides followed by the base –hedron, known generically as *n*-hedrons, similar to *n*-gons. Since the number and complexity of polyhedra are unlimited, the names can become complex also. The table below shows the names and number of sides for some of the most common polyhedron shapes.

Name	Faces	Name	Faces	Name	Faces
Tetrahedron	4	Nonahedron	9	Icosahedron	20
Pentahedron	5	Decahedron	10	Icositetrahedron	24
Hexahedron	6	Undecahedron	11	Tricontahedron	30
Heptahedron	7	Dodecahedron	12	Icosidodecahedron	32
Octahedron	8	Tetradecahedron	14	Hexecontrahedron	60

Table 6-1: Polyhedra Names and Faces

6-3 Polyhedron Name – Number of Faces

Naming a polyhedron using only the number of faces is usually not sufficient to identify a specific polyhedron. The following example will help to illustrate this concept.

A polyhedron with six sides is called a hexahedron. There are seven distinct hexahedra: triangular dipyramid, pentagonal pyramid, tetragonal antiwedge, hemiobelisk, hemicube, cube, and pentagonal wedge. The cube, or regular hexahedron, is the most common. The table below shows the number of faces, edges, and vertices for each of the seven hexahedra.

Name of Hexahedron	Faces	Edges	Vertices
Triangular Dipyramid	6	9	5
Pentagonal Pyramid	6	10	6
Tetragonal Antiwedge	6	10	6
Hemiobelisk	6	11	7
Hemicube	6	11	7
Cube	6	12	8
Pentagonal Wedge	6	12	8

Table 6-2: Polyhedra – Faces, Edges, Vertices

A triangular dipyramid is the dual of a triangular prism, and looks like two tetrahedra glued on a common face. A pentagonal pyramid is a pyramid whose base is a pentagon. Like all pyramids, it is self-dual. A tetragonal antiwedge is a skewed pentagonal pyramid and is the least symmetric of the hexahedra. Its dual is a shape of its mirror own image. A hemiobelisk is an elongated square pyramid or obelisk with one of the four base corners of the square pyramid cut off to create a new triangular face. A hemicube is a cube with a plane cutting through two opposite corners and the midpoint of two edges. A cube is a regular hexahedron. A pentagonal wedge is a tetrahedron with two corners cut off. The figures below show the seven hexahedra.

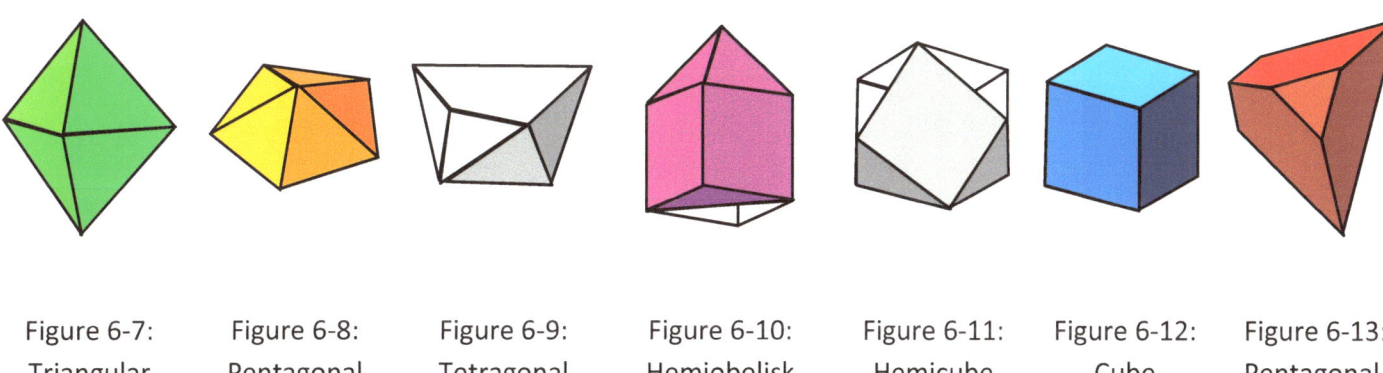

Figure 6-7: Triangular Dipyramid

Figure 6-8: Pentagonal Pyramid

Figure 6-9: Tetragonal Antiwedge

Figure 6-10: Hemiobelisk

Figure 6-11: Hemicube

Figure 6-12: Cube

Figure 6-13: Pentagonal Wedge

6-4 Polyhedron Name – Shape of Faces

Naming a polyhedron using only the shape of the faces is usually not sufficient to identify a specific polyhedron. The following example will help to illustrate this concept.

A polyhedron whose faces are all equilateral triangle is called a ==deltahedron==. A deltahedron has an even number of faces (2n faces, 3n edges, and n+2 vertices). There are eight convex deltahedra: ==regular tetrahedron==, ==triangular dipyramid==, ==regular octahedron== or square dipyramid, ==pentagonal dipyramid==, ==snub disphenoid==, ==triaugmented triangular prism==, ==gyroelongated square dipryamid==, and ==regular icosahedron== or gyroelongated pentagonal dipyramid. The table below shows the number of faces, edges, and vertices for each of the eight deltahedra.

Name of Deltahedron	Faces	Edges	Vertices
Regular Tetrahedron	4	6	4
Triangular Dipyramid	6	9	5
Regular Octahedron	8	12	6
Pentagonal Dipyramid	10	15	7
Snub Disphenoid	12	18	8
Triaugmented Triangular Prism	14	21	9
Gyroelongated Square Dipyramid	16	24	10
Regular Icosahedron	20	30	12

Table 6-3: Deltahedra – Faces, Edges, Vertices

Three of the deltahedrons are Platonic solids: regular tetrahedron, regular octahedron, and regular icosahedron. Three of the deltahedrons are dipyramids: triangular dipyramid, square dipyramid, and pentagonal dipyramid. Three of the deltahedrons are Johnson solids: snub disphenoid, triaugmented triangular prism, and gyroelongated square dipyramid. The octahedron, as a Platonic solid, and the square dipyramid, as a dipyramid, are the same polyhedron and counted twice. The figures below show the eight deltahedra.

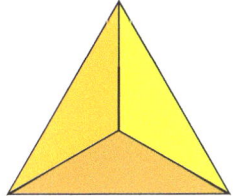
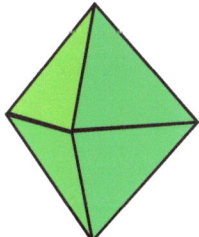
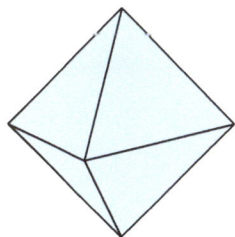
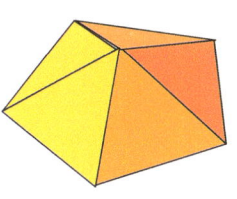

Figure 6-14: Regular Tetrahedron

Figure 6-15: Triangular Dipyramid

Figure 6-16: Regular Octahedron

Figure 6-17: Pentagonal Dipyramid

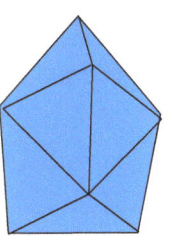
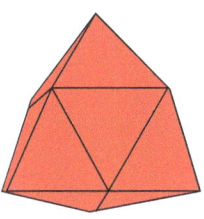
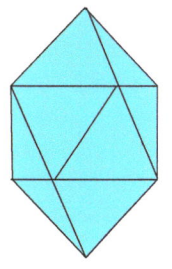
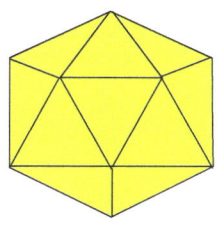

Figure 6-18: Snub Disphenoid

Figure 6-19: Triaugmented Triangular Prism

Figure 6-20: Gyroelongated Square Dipyramid

Figure 6-21: Regular Icosahedron

6-5 Polyhedron Symmetry

Symmetry is a concept of balance and self-similarity. A polyhedron can be rotated around its center point along the X, Y, and Z axis. After a rotation, a symmetrical polyhedron should appear similar to the original view. For example, one side of a polyhedron looks like a hexahedron. If after a rotation, the opposite side can be seen and it also looks like a hexahedron, then the polyhedron is symmetrical. However, if the opposite side looks like a square pyramid, then the polyhedron is not symmetrical. The rotation around the X axis is seen as an object turning in a clockwise or counterclockwise direction. The rotation around the Y axis is seen as an object turning the top closer or the top farther. The rotation around the Z axis is seen as an object spinning to the left or right. The figures below show a polyhedron rotating around the X, Y, and Z axis.

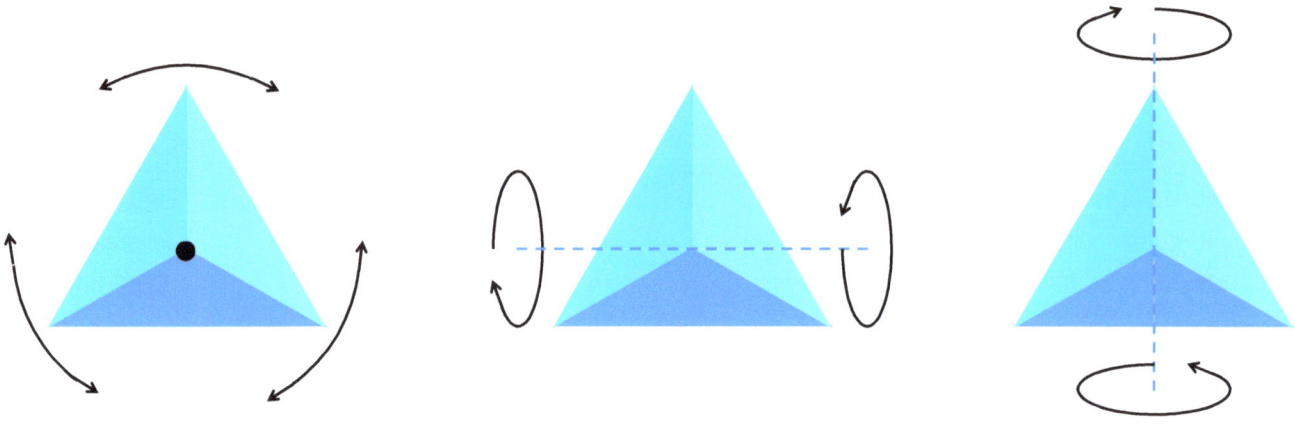

Figure 6-22: X Axis Rotation   Figure 6-23: Y Axis Rotation   Figure 6-24: Z Axis Rotation

Most polyhedra are highly symmetrical and within a symmetric orbit. The symmetry orbit of a polyhedron refers to its circumscribed sphere. The circumscribed sphere is the three-dimensional analogue of the circumscribed circle. A circumscribed sphere of a polyhedron is a sphere that contains the polyhedron and touches each of the polyhedron's vertices. All regular or semiregular polyhedra have circumscribed spheres, but most irregular polyhedra do not have one, since in general not all vertices lie on a common sphere. The radius of a sphere circumscribed around a polyhedron P is called the circumradius of P. A two-dimensional square can rotate around its center within its circumscribed circle. A three-dimensional octahedron can rotate around its center within its circumscribed sphere. The figures below show a circumscribed circle and a circumscribed sphere.

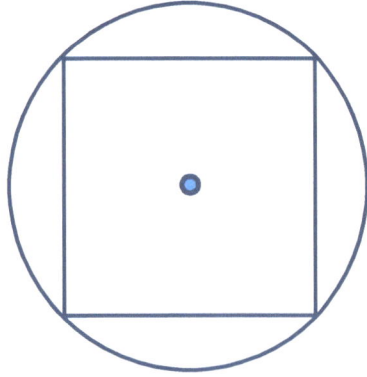 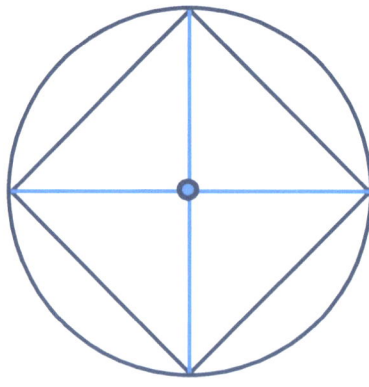

Figure 6-25: Circumscribed Circle   Figure 6-26: Circumscribed Sphere

6-6 Polyhedron Characteristics

A polyhedron can be classified according to its characteristics. The type of symmetry of a polyhedron is based on the similarity of the vertices, edges, and faces. An isogonal or vertex-transitive polyhedron has symmetrical vertices. This means that each vertex is surrounded by the same kinds of face in the same or reverse order and with the same angles between the corresponding faces. An isotoxal or edge-transitive polyhedron has symmetrical edges. This means that there is only one type of edge to an object, such as a hexagon face meeting another hexagon face. The dihedral angle, or angle at which two faces meet, is the same for all edges. An isohedral or face-transitive polyhedron has symmetrical faces. This means that all of the faces must be congruent.

Polyhedra are classified into categories based on their isogonal, isotoxal, and isohedral symmetrical properties. The six commonly used categories are regular, quasi-regular, semi-regular, uniform, non-uniform, and noble. A polyhedron can belong to more than one group. The table below shows polyhedron categories.

Category	Properties	Solids Group	Number
Regular	Isogonal, Isotoxal, Isohedral, every face is the same regular polygon	Platonic, Kepler-Poinsot	5, 4
Quasi-regular	Isogonal, Isotoxal, every face is regular polygon	Archimedean – only Cuboctahedron and Icosidodecahedron	2
Semi-regular	Isogonal, every face is regular polygon	Archimedean	13
Uniform	Regular, Quasi-regular, or Semi-regular	----------	--
Non-uniform	Not isogonal, every face is regular polygon	Johnson	92
Noble	Isogonal, Isohedral	Platonic, others	--

Table 6-4: Polyhedra Properties

Every polyhedron is uniquely related to another specific polyhedron called its dual polyhedron. The polyhedron and the dual polyhedron have the same number of edges, but the vertices and faces are reversed or occupy complementary locations. The dual polyhedron of a dual polyhedron is the original polyhedron. The dual of a polyhedron is created by connecting the center of each face with a line to form the edges of each face. The following figures show the dual process to transform a cube to an octahedron and an octahedron to a cube.

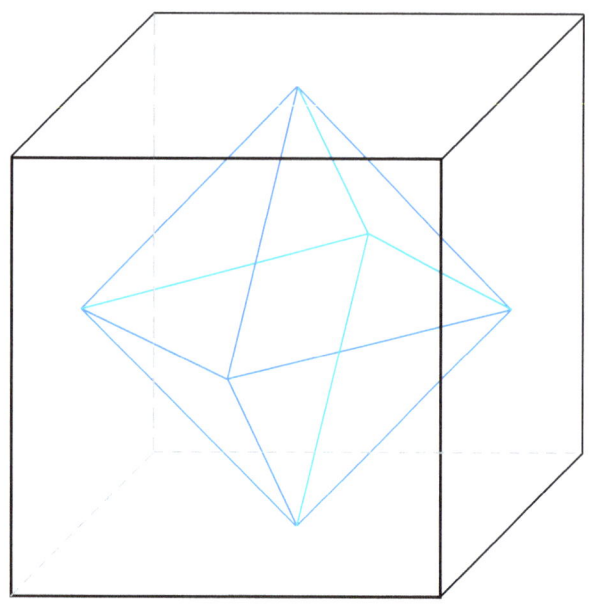
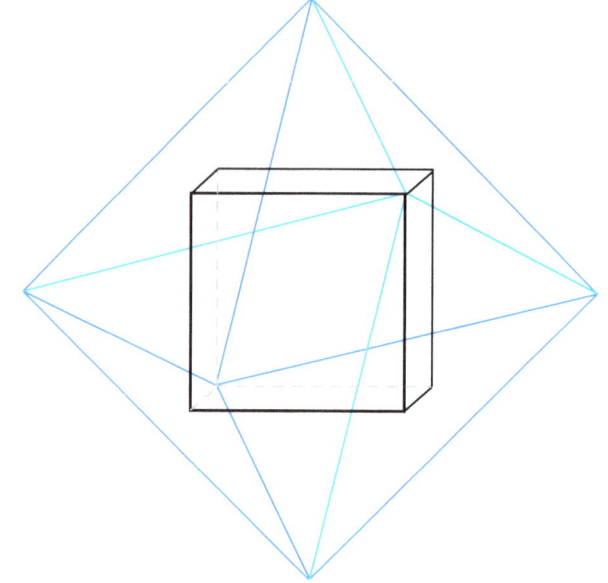

Figure 6-27: Cube to Octahedron   Figure 6-28: Octahedron to Cube

A polyhedron can be <mark>self-dual</mark>, such as the tetrahedron. This means that the dual polyhedron of a tetrahedron is a tetrahedron. In some instances, an uncommon polyhedron may have a dual polyhedron, but that dual polyhedron does not have a widely accepted name. The table below shows the dual polyhedron for the Platonic, Kepler-Poinsot, and Archimedean groups of polyhedron.

Group	Polyhedron	Dual Polyhedron
Regular – Platonic	Tetrahedron	Tetrahedron
	Cube	Octahedron
	Octahedron	Cube
	Dodecahedron	Icosahedron
	Icosahedron	Dodecahedron
Regular – Kepler-Poinsot	Small Stellated Dodecahedron	Great Dodecahedron
	Great Dodecahedron	Small Stellated Dodecahedron
	Great Stellated Dodecahedron	Great Icosahedron
	Great Icosahedron	Great Stellated Dodecahedron
Archimedean	Cuboctahedron	Rhombic Dodecahedron
	Icosidodecahedron	Rhombic Triacontahedron
	Truncated Tetrahedron	Triakis Tetrahedron
	Truncated Octahedron	Tetrakis Hexahedron
	Truncated Cube	Triakis Octahedron
	Truncated Icosahedron	Pentakis Dodecahedron
	Truncated Dodecahedron	Triakis Icosahedron
	Rhombicuboctahedron	Deltoidal Icositetrahedron
	Truncated Cuboctahedron	Disdyakis Dodecahedron

	Rhombicosidodecahedron	Strombic Hexecontahedron
	Truncated Icosidodecahedron	Disdyakis Triacontahedron
	Snub Cuboctahedron	Petaloidal Disdodecahedron
	Snub Dodecahedron	Pentagonal Hexecontahedron

Table 6-5: Dual Polyhedra

Kepler-Poinsot solids are regular but not convex. A list of all convex polyhedra whose faces are regular polygons includes 5 Platonic solids, 13 Archimedean solids, and 92 Johnson solids. There are also infinitely many prisms and antiprisms.

6-7 Euler's Formula

Euler's polyhedron formula defines the number of faces, edges, and vertices of a spherical polyhedron. The 18th-century Swiss mathematician Leonhard Euler showed that for any simple convex polyhedron, the sum of the number of vertices V and the number of faces F is equal to the number of edges E plus 2, or $V+F=E+2$. This formula can be used for all Platonic, Archimedean, and Johnson solids, but not for Kepler-Poinsot solids. This is a useful formula for determining the number of each vertex, face, and edge of a polyhedron. The table below shows the number of faces, edges, and vertices for each of the five Platonic solids.

Name of Platonic Solid	Faces	Edges	Vertices	Euler's Formula
Tetrahedron	4	6	4	$4 + 4 = 6 + 2$
Cube	6	12	8	$8 + 6 = 12 + 2$
Octahedron	8	12	6	$6 + 8 = 12 + 2$
Dodecahedron	12	30	20	$20 + 12 = 30 + 2$
Icosahedron	20	30	12	$12 + 20 = 30 + 2$

Table 6-6: Platonic Solids – Faces, Edges, Vertices

6-8 Polyhedron Nets

A polyhedron can be unfolded along some of its edges until its surface is spread flat like a sheet of paper. The resulting flat diagram is called a net. A net contains all the faces of a polyhedron, with some of them separated by angular gaps. A net is a flat pattern that can then be folded along the edges and taped together to regenerate the polyhedron. A net enables the easy construction of a basic polyhedron out of paper. The construction of net models can help to illustrate the volume and surface areas of a polyhedron, making concepts in geometry easier to learn. The figures below show five polyhedra and their corresponding nets. The shaded face in each net diagram corresponds to the base of each polyhedron.

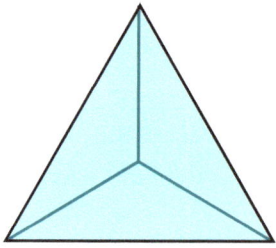
Figure 6-29: Tetrahedron

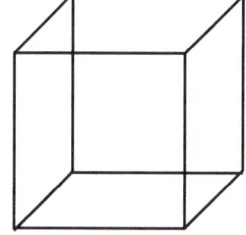
Figure 6-30: Cube

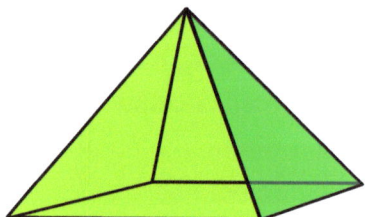
Figure 6-31: Square Pyramid

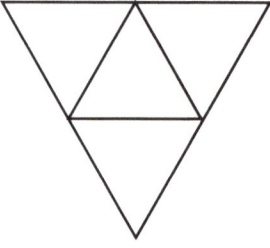
Figure 6-32: Tetrahedron Net

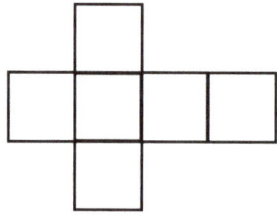
Figure 6-33: Cube Net

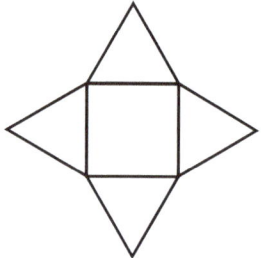
Figure 6-34: Square Pyramid Net

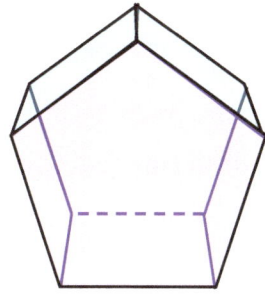

Figure 6-35: Pentagonal Prism

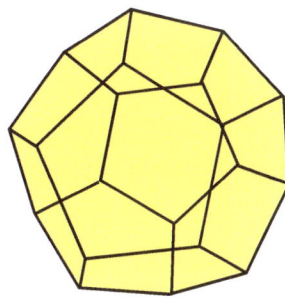

Figure 6-36: Dodecahedron

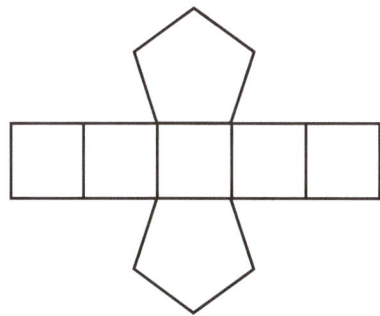

Figure 6-37: Pentagonal Prism Net

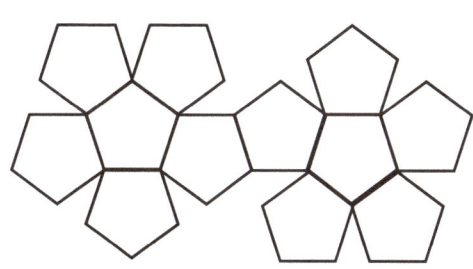

Figure 6-38: Dodecahedron Net

• 6-9 Summary

A polyhedron is a three-dimensional shape bounded by faces (polygons), edges (line segments), and vertices (points). Since a polyhedron is comprised of polygons, it has flat faces and straight edges. A polyhedron is an enclosed three-dimensional object. There are three definitions of a polyhedron based on bounding characteristics. The first describes a polyhedron as only the edges of the polygons. There is no surface area or internal volume. It resembles a wire frame. The second states a polyhedron is a boundary between the internal and external areas formed by the faces of polygons. The volume of the internal space is not included. It is a hollow shape. The third includes the polygon surfaces and the internal volume. A polyhedron is a three-dimensional solid. When describing a polyhedron in general, the third definition is almost always used.

The name of a polyhedron can be based on several factors, such as the number of faces, the polygon shape of the faces, and the convex or concave surface of the faces. A polyhedron is convex if its surface does not intersect itself and a line segment drawn to connect any two points of the polyhedron is contained in the interior or on the surface. A polyhedron is concave if it has a hole or indentation and is complex if the surface is not uniformly flat or is like a star. Naming a polyhedron using only the number of faces or using only the shape of the faces is usually not sufficient to identify a specific polyhedron.

Symmetry is a concept of balance and self-similarity. A polyhedron can be rotated around its center point along the X, Y, and Z axis. After a rotation, a symmetrical polyhedron should appear similar to the original view. The rotation around the X axis is seen as an object turning in a clockwise or counterclockwise direction. The rotation around the Y axis is seen as an object turning the top closer or the top farther. The rotation around the Z axis is seen as an object spinning to the left or right. Most polyhedra are highly symmetrical and within a symmetric orbit. The symmetry orbit of a polyhedron refers to its circumscribed sphere. The circumscribed sphere is the three-dimensional analogue of the circumscribed circle. A circumscribed sphere of a polyhedron is a sphere that contains the polyhedron and touches each of the polyhedron's vertices.

A polyhedron can be classified according to its characteristics. The type of symmetry of a polyhedron is based on the similarity of the vertices, edges, and faces. An isogonal or vertex-transitive polyhedron has symmetrical vertices. This means that each vertex is surrounded by the same kinds of face in the same or reverse order and with the same angles between the corresponding faces. An isotoxal or edge-transitive polyhedron has symmetrical edges. This means that there is only one type of edge to an object, such as a hexagon face meeting another hexagon face. The dihedral angle, or angle at which two faces meet, is the same for all edges. An isohedral or face-transitive polyhedron has symmetrical faces. This means that all of the faces must be congruent.

Every polyhedron is uniquely related to another specific polyhedron called its dual polyhedron. The polyhedron and the dual polyhedron have the same number of edges, but the vertices and faces are reversed or occupy complementary locations.

Euler's polyhedron formula defines the number of faces, edges, and vertices of a spherical polyhedron. The 18th-century Swiss mathematician Leonhard Euler showed that for any simple convex polyhedron, the sum of the number of vertices V and the number of faces F is equal to the number of edges E plus 2, or $V+F=E+2$. This formula can be used for all Platonic, Archimedean, and Johnson solids, but not for Kepler-Poinsot solids. This is a useful formula for determining the number of each vertex, face, and edge of a polyhedron.

A polyhedron can be unfolded along some of its edges until its surface is spread flat like a sheet of paper. The resulting flat diagram is called a net. A net contains all faces of a polyhedron, with some of them separated by angular gaps. The construction of net models can help to illustrate the volume and surface areas of a polyhedron, making concepts in geometry easier to learn.

CHAPTER 6

Chapter Test

Grading Scale: One point for each correct answer.

Excellent = 43-47, Good = 38-42, Average = 33-37, Fair = 29-32, Poor = 0-28

• 6-1 Introduction

Mark as True or False.

1. A polyhedron is a three-dimensional shape bounded by faces, edges, and vertices. ____
2. Every side of every polygon belongs to more than one other polygon. ____
3. The faces that share a vertex form a chain of polygons, but do not share a side. ____
4. A polyhedron is an enclosed three-dimensional object. ____
5. A polyhedron can resemble a wire frame, a hollow shape, or a solid shape. ____

• 6-2 Polyhedron Name Conventions

Match the number of faces and the polyhedron.

A = 4   B = 5   C = 8   D = 9   E = 20   F = 60

1. Icosahedron ____
2. Pentahedron ____
3. Nonahedron ____
4. Hexecontrahedron ____
5. Tetrahedron ____
6. Octahedron ____

• 6-3 Polyhedron Name – Number of Faces

Match the definitions and terms.

A = Triangular Dipyramid   B = Pentagonal Pyramid   C = Tetragonal Antiwedge
D = Hemiobelisk   E = Hemicube   F = Cube
G = Pentigonal Wedge   H = Hexahedron

1. Elongated square pyramid or obelisk with one of the base corners cut off. ____
2. Pyramid whose base is a pentagon. ____

71

3. Tetrahedron with two corners cut off.
4. Cube with a plane cutting two opposite corners and the midpoint of two edges.
5. Polyhedron with six sides.
6. Dual of a triangular prism, and like two tetrahedra glued on a common face.
7. Skewed pentagonal pyramid and is the least symmetric of the hexahedra.
8. Regular hexahedron.

6-4 Polyhedron Name – Shape of Faces

Match the definitions and terms.

A = Regular Tetrahedron B = Triangular Di pyramid C = Regular Octahedron
D = Pentagonal Dipyramid E = Snub Disphenoid
F = Triaugmented Triangular Prism G = Gyroelongated Square Dipyramid
H = Regular Icosahedron

1. All equilateral triangle faces with faces = 20, edges = 30, and vertices = 12.
2. All equilateral triangle faces with faces = 6, edges = 9, and vertices = 5.
3. All equilateral triangle faces with faces = 10, edges = 15, and vertices = 7.
4. All equilateral triangle faces with faces = 14, edges = 21, and vertices = 9.
5. All equilateral triangle faces with faces = 12, edges = 18, and vertices = 8.
6. All equilateral triangle faces with faces = 8, edges = 12, and vertices = 6.
7. All equilateral triangle faces with faces = 16, edges = 24, and vertices = 10.
8. All equilateral triangle faces with faces = 4, edges = 6, and vertices = 4.

6-5 Polyhedron Symmetry

Match definitions and terms.

A = Symmetry B = X-axis Rotation C = Y-axis Rotation
D = Z-axis Rotation E = Circumscribed Sphere

1. An object turning the top closer or the top farther.
2. An object spinning to the left or right.
3. An object turning in a clockwise or counterclockwise direction.
4. A sphere that contains the polyhedron and touches each of the vertices.
5. A concept of balance and self-similarity.

6-6 Polyhedron Characteristics

Match definitions and terms.

    A = Isogonal   B = Isotoxal   C = Isohedral
    D = Dihedral Angle   E = Dual Polyhedron

1. Polyhedron with symmetrical edges. _____
2. Polyhedron with the same number of edges, but vertices and faces are reversed. _____
3. Polyhedron with symmetrical faces. _____
4. Polyhedron with symmetrical vertices. _____
5. The angle at which two faces meet. _____

6-7 Euler's Formula

Calculate the number of vertices, faces, or edges of a polyhedron. The Euler's formula is as follows: $V + F = E + 2$.

1. Vertices for Cuboctahedron with Faces = 14 and Edges = 24 _____
2. Faces for Rhombicubcahedron with Edges = 48 and Vertices = 24 _____
3. Edges for Snub Dodecahedron with Vertices = 60 and Faces = 92 _____
4. Vertices for Triakis Tetrahedron with Faces = 12 and Edges = 18 _____
5. Faces for Deltoidal Icositetrahedron with Edges = 48 and Vertices = 26 _____
6. Edges for Pentagonal Hexecontahedron with Vertices = 92 and Faces = 60 _____

6-8 Polyhedron Nets

Mark as True or False.

1. A three-dimensional model of a polyhedron is called a net. _____
2. A net can not be used to construction of a basic polyhedron out of paper. _____
3. Net models can help to illustrate the volume and surface areas of a polyhedron. _____
4. Net models make concepts in geometry easier to learn. _____

www.ingramcontent.com/pod-product-compliance
Lightning Source LLC
Chambersburg PA
CBHW051024180526
45172CB00002B/459